SPEED MATHEMATICS SIMPLIFIED

Edward Stoddard

DOVER PUBLICATIONS, INC.
New York

Bibliographical Note

This Dover edition, first published in 1994, is an unabridged and unaltered republication of the second printing (1965) of the work first published by The Dial Press, New York, in 1962.

Library of Congress Cataloging-in-Publication Data

Stoddard, Edward.
Speed mathematics simplified / by Edward Stoddard. — Dover ed.
p. cm.
Originally published: New York : Dial, 1962.
Includes bibliographical references.
ISBN 0-486-27887-5
1. Ready-reckoners. I. Title.
QA111.S85 1994
513′.9—dc20 93-46724
CIP

Manufactured in the United States of America
Dover Publications, Inc., 31 East 2nd Street, Mineola, N.Y. 11501

CONTENTS

INTRODUCTION

WHETHER you are an executive concerned with inventories and markups and profit ratios or a carpenter who works with board feet and squares of shingles—whether you do your figuring in gallons and pennies or tons and dollars—this book will show you new ways to do that figuring with dispatch and authority.

With the techniques in this book, you will find yourself doing many problems in your head that formerly required pencil and paper. More complex problems that still need pencil and paper will get done in a fraction of the former time, and in many cases you will simply jot down two or three numbers rather than copy down the whole problem.

When a quick estimate or accurate guess is needed, you will be the one who can glance at a column of figures or a complicated multiplication and give a rapid approximation accurate to any number of places needed.

If all this sounds too good to be true, let me hasten to point out that there are some things this book *cannot* do:

This book cannot make you a "number genius" who multiplies a six-digit number by a twelve-digit number in his head and gives the complete answer in ten seconds flat. There are such people, but they are born—not trained. There are mighty few of them, at that.

This book cannot hand you mastery of streamlined math on a silver platter. It can show you the techniques, explain each of them as clearly and simply as possible, and encourage you to do the pleasantest possible kind of practice. But only you can decide to spend the necessary time the explanations and the practice will inevitably take.

You have already taken the first major step in mastering speed math. You bought or borrowed this book because you want to become better at figures. Wanting to learn is basic. If your interest ever flags, if the practice ever seems irksome, it might be well to remind yourself why you picked up the book in the first place. Keeping the goal in mind is the best way to keep your feet firmly on the path.

There are at least half a dozen books in print on "speed" or "short-cut" mathematics.

Why, then, this one?

There are a number of good reasons. First, almost all books on the subject rely primarily on a number of standard short cuts. The use of these devices, which include such simple conversions as aliquot parts and factoring, can often save a great deal of time. As far as I have been able to find out, however, no book has yet attempted to relate them to each other and show the ways to pick out the most useful in each case. Here you will find the most valuable of the classic short cuts explained quite simply and arrranged for sensible, rapid selection and use.

Beyond this, the book introduces an entirely new system of basic figuring that works in all cases. This approach builds on the arithmetic you already know. It takes your present training in numbers and streamlines it, cutting down the number of steps you take in solving each problem. By combining this approach with the best of the classic short cuts, you will compound your speed and ease.

This new system is a development of a little-known oriental technique growing directly out of abacus theory. The abacus is a startlingly efficient machine, for all the jokes made about it, mainly because it forced on the orientals who perfected the modern version a simplified approach to numbers.

The chapter on addition will go more fully into the contributions of the abacus to this system.

One more point about this book. Simply reading through

it will accomplish little. Practice is required to master any activity, whether it be streamlined mathematics or water skiing. I have already mentioned the importance of this, but very few of us have the patience to work out small-print examples or the self-control to avoid peeking at the answers printed right beside them.

That is why you will find a different method of practice here. It bears some similarity to the new theories of "teaching machines" in that it requires you to produce the answer and, immediately after, tells you whether you were right or wrong. In addition, I have kept the practice as varied as possible, and tried to give it some pace as well. The method is designed to give you enough basic practice as you go to begin mastery of each step.

Please do not skip these sections. They are absolutely essential to learning how to *use* streamlined math. They carry you from knowing how it is done to knowing how to do it—quite a different thing, really.

This is how these sections work:

As you come to an example or series of speed-practice steps, you will be asked to cover the answer (if it is on the same page) with a bit of working paper you should always keep on hand. Use the paper for any pencil figuring involved. I would recommend that you tuck a dozen blank file cards into the book for this purpose, or a thin pad no larger than the book. It can serve the additional use of a bookmark, too. A good idea would be to stop for a moment and get hold of a pencil and pad or cards right now.

When you come to a demonstration or practice problem, read it. Be sure you understand the specific technique to be used. Then work it out, keeping your paper over the answer. If a pencil is needed, work it out on the paper. Then, and only then, look at the answer. If you made a mistake, stop to see why before going on.

Do this faithfully if you want to get all the good from this book.

As in learning any new skill, you may feet a bit awkward and slow at first. This is entirely natural. Repetition and time will cure the awkwardness. The only way to learn to ice-skate is to ice-skate. The only way to learn speed mathematics is to use (not merely read about) speed mathematics.

By the time you have finished this book, your speed and ease with figures should easily have doubled. From then on, as you make these techniques automatic and habitual, your skill will continue to improve. You can ensure this in two ways:

First, consciously use the new ways you have learned for every number problem you run across in business or personal life. At the beginning you will have to strain a bit to break the old habits, and the process will take a little longer because it is new. But soon you will find yourself using these techniques comfortably and quickly. As you continue using them, you will find yourself approaching any number in this new way without even thinking about it.

Second, do a bit of special practice now and then just for fun. Instead of doing a crossword puzzle on the train, run through a few random problems using your new techniques. Instead of reading an old magazine while waiting for an appointment, do some mental exercising with the phone number or street address of the office where you are waiting. Instead of killing half an hour with a TV program you don't especially want to look at anyway, go through one of the speed-developer chapters in this book again.

Do all of these things cheerfully and conscientiously, making a game of them, and with only a reasonable amount of time and patience you will find yourself becoming truly a whiz at figures.

SPEED MATHEMATICS SIMPLIFIED

1

NUMBER SENSE

NUMBER sense is our name for a "feel" for figures—an ability to sense relationships and to visualize completely and clearly that numbers only symbolize real situations. They have no life of their own, except as a game.

Almost all of us disliked arithmetic in school. Most of us still find it a chore.

There are two main reasons for this. One is that we were usually taught the hardest, slowest way to do problems because it was the easiest way to teach. The other is that numbers often seem utterly cold, impersonal, and foreign.

W. W. Sawyer expresses it this way in his book *Mathematician's Delight:* "The fear of mathematics is a tradition handed down from days when the majority of teachers knew little about human nature, and nothing at all about the nature of mathematics itself. What they did teach was an imitation."

By "imitation," Mr. Sawyer means the parrot repetition of rules, the memorizing of addition tables or multiplication tables without understanding of the simple truths behind them.

Actually, of course, in real life we are never faced with an abstract number four. We always deal with four tomatoes, or four cats, or four dollars. It is only in order to learn how to deal conveniently with the tomatoes or the cats or the dollars that we *practice* with an abstract four.

In recent years, teachers of mathematics have begun to express concern about popular understanding of numbers. Some advances have been made, especially in the teaching of fractions by diagrams and by colored bars of different lengths to help students visualize the relationships.

About the problem-solving methods, however, very little has been done. Most teaching is of methods directly contrary to speed and ease with numbers.

When I coached my son in his multiplication tables a year ago, for instance, I was horrified at the way he had been instructed to recite them. I had made up some flash cards and was trying to train him to "see only the answer"—a basic technique in speed mathematics explained in the next few pages. He hesitated, obviously ill at ease. Finally he blurted out the trouble:

"They don't let me do it that way in school, Daddy," he said. "I'm not allowed to look at 6×7 and just say '42.' I have to say 'six times seven is forty-two.' "

It is to be hoped that this will change soon—no fewer than three separate professional groups of mathematics teachers are re-examining current teaching methods—but meanwhile, we who went through this method of learning have to start from where we are.

Relationships

Even though arithmetic is basically useful only to serve us in dealing with solid objects, be they stocks, cows, column inches, or kilowatts, the fact that the same basic number system applies to all these things makes it possible to isolate "number" from "thing."

This is both the beauty and—to schoolboys, at least—the terror of arithmetic. In order fully to grasp its *entire* application, we study it as a thing apart.

For practice purposes, at least, we forget about the tomatoes and think of the abstract concept "4" as if it had a real existence of its own. It exists at all, of course, only in the method of thinking about the tools we call "numbers" that we have slowly and painstakingly built up through thousands of years.

There is space here only to touch briefly on the intriguing results of the fact that we were born with ten fingers, and therefore use ten as a base number for our entire counting system. Other systems have been and still are used, from the binary system based on two required by digital electronic computers to the duo-decimal (dozens) base still in use in buying eggs, products by the gross, English money, inches to the foot, and hours to the day.

Our counting system is based on 10, because we have 10 fingers. As refined and perfected over the centuries, it is a wonderful system.

Everything you ever need to do in arithmetic, whether it happens to be calculating the concrete to go into a dam or making sure you aren't overcharged on a three-and-a-half pound chicken at 49½¢ a pound, can and will be done within the framework of ten.

A surprisingly helpful exercise in feeling relationships of the numbers that go into ten is to spend a few moments with the following little example.

First, look at these three dots:

• • •

Nothing very exciting yet. But now we add three more dots, right below them:

• • •
• • •

How many dots are there? Six, of course. But how did it come about that there are now six? We added three dots to the first three. Then what is three plus three?

Of course you know the answer, and of course this seems pedestrian. But there is a moral.

Did we also double the first number of dots? There were three, and we added the same number. Now there are six. So what is three plus three, again? And what is two times three?

You know the answer, but sit back for a moment and try to visualize the six dots. They are both three plus three, and two times three. The better emotional grasp of this you can get now, the more firmly you can feel as well as understand this relationship, the faster and easier the rest of the book will go.

Now we add three more dots:

• • •

• • •

• • •

How many dots?

What is three times three? Can you feel it? What is six plus three? Pause as you answer to let it sink in.

What is one-third of nine?

Play with these dots a bit. Try to see as many relationships as you can. Notice that three-ninths is equal to one-third. Why? What is six-ninths in simpler numbers?

Oddly enough, all of our arithmetic—even into the millions—is based on the number of dots you now have in front of you. Add one to nine and you have ten—which is the base of our counting system. We express it with a new one moved over to mean *one ten* and a zero to mean nothing—nothing more than ten.

If we really have a feel for all the relationships within the number nine, we are a long way toward feeling at home with numbers.

Stop for a bit here and, on your pad, set up ten dots. Amuse yourself by setting them up in two rows of five each. See what happens if you try to make any other number of rows with the same number of dots in each row come out to ten. Look back at the two rows of five each and see if you can feel the reason why we can express one-fifth and one-half of ten (or one) with a single-digit decimal, but not one-third or one-fourth.

Seeing Only the Answer

Beyond working at a "feel" for number relationships there are certain specific rules of procedure that will speed up your handling of numbers.

The first of these is simply a matter of training. Quite new training for many of us, and one directly contrary to the way arithmetic is often taught, but one that offers an amazing improvement all by itself.

The technique is to see only the answer.

When adding, we learn to "see" the two digits 4 and 3 as 7—not as 4 and 3.

Then, multiplying, we learn to "see" the digits 4 and 3 as 12—not as 4 and 3.

This may seem elementary. You may already be doing something very much like it in your own number handling. Yet some conscious work in this direction will pay handsome dividends.

Try to remember, if you can, how it was when you first learned to read. You spelled out each word letter by letter. It was slow and painful and not really very enjoyable. But now you grasp whole words and phrases at a glance. It's not only faster, it is easier.

This is unfortunately just the opposite to the way most arithmetic is taught, so most of us have to unlearn what was drilled into us in school. But it is well worth the effort, and it is essential to many of the streamlined methods and short cuts later in the book.

Arithmetic has been called the language of business. In many most important senses it really is, and in order to understand income-expense and financial statements you need a good grasp of it. Our insistence on the importance of seeing only the answer—of seeing 6×7 as 42—is basic to a vocabulary of the language. The methods and short cuts to come later might be called the grammar, but grammar is useless without vocabulary.

From time to time in this book I will slip in a little casual practice at seeing only the answer. Please do not skip these examples. They are important. They directly affect every other element in the book.

Add these numbers: 8 7 6

Did you see the digits 8, 7, and 6? You were probably taught to add "8 plus 7 is 15; 15 plus 6 is 21."

This is too slow.

Instead, practice looking at the 8 and the 7 and thinking, automatically, "15." Try to do this without saying or thinking either the 8 or the 7. Then, thinking only "15," glance at the 6 and *see* "21." You don't say or even think "6" at all.

If you have never tried this, the idea may be not only new but rather shocking. You can get used to it very quickly if you try, and it will speed up your number work substantially even without the other techniques. It isn't hard. It takes a bit of practice, and knowing your addition tables so you don't have to cudgel your brains to remember what 8 and 7 add up to. It's just what you do when you look at *m* and *e* and think "me" without consciously putting the two letters together.

Try it again: 8 7 6

Now practice reading the following additions by seeing only the answer. Don't say to yourself, and try to avoid even thinking to yourself, the digits you are adding. Do your best to "see" 4 plus 5 as 9—not as 4 plus 5. Read the answers to these additions just as you would read *i* and *t* as *it,* not *i* and *t*:

$$\begin{array}{cccc} 1 & 3 & 6 & 5 \\ \underline{4} & \underline{2} & \underline{2} & \underline{4} \\ \\ 3 & 4 & 2 & 8 \\ \underline{7} & \underline{2} & \underline{5} & \underline{1} \\ \\ 9 & 6 & 1 & 3 \\ \underline{1} & \underline{4} & \underline{1} & \underline{4} \end{array}$$

If you found yourself beginning to see only the answers, very good. If not, you might find it helpful to try again.

Work With Numbers, Not Digits

The second step to developing number sense goes even further in aiding a natural and sure speed with figures. This step is far more drastic than seeing only the answer. It violates almost everything we are usually taught about numbers, yet you will quickly see how much sense it makes and how important it can be.

This rule, agreed on by almost every teacher of short-cut mathematics, is to work from left to right—not right to left.

This is just opposite to what is taught in school. We are taught to add, subtract, and multiply from right to left. It is easier to teach to children and easier to learn from the "imitation" standpoint of learning by rote, but it is directly contrary to the way we read and think about numbers.

There are at least three important advantages to working from left to right.

First, it is the way we look at everything else on a page. We read from left to right.

Second, it is the way we look at a number right up to the moment we begin doing something to it. For instance, look at the number 164,928. You *read* it one-hundred sixty-four thousand, nine-hundred and twenty-eight. But when you begin to add or subtract or multiply it, you are taught to tackle it as 8, 2, 9, 4, 6, 1.

It isn't the same number at all. At the very outset we are taught to combine this number with another in a totally foreign, unrecognizable form.

There is still a third reason why it is faster and better to work from left to right. You develop your most important numbers first and work toward the "details."

Suppose you are a salesman who has just sold a $423 order for which you will get a 6% commission. If you work from left to right (you will learn how later), you know by the time you get just one digit that your commission will be twenty-something dollars. You know when you have finished two digits that your commission is $25 and change.

But if you work in the schoolroom, right-to-left way, the first two digits you develop tell you only the change. You know only that you will get something dollars and 38¢. Not until you finish working out the whole commission do you know that your commission will be $25.38.

That 38¢ may be important to a bookkeeper, but its importance in the number itself is relatively a detail. You care a lot more about the $25 than you do about the 38¢.

This is true of every number and every application, whether or not a decimal point happens to break it into dollars and cents. The first digit in a number is *ten times* as important as the second, a *hundred times* as important as the third, and so on down the line. If the order we just discussed were a hundred times as large, you would still care a great deal more about the $2,500 part of the commission than you would about the $38 part.

Working from left to right reveals to you, step by step, the most important numbers first. For this reason alone, the new methods for doing this are one of the most valuable quick estimating tools you can have.

The fact that each digit in a number decreases in importance by a factor of ten as it moves one place to the right is the reason why many companies today report their operations and financial position in "round" numbers: rounding off the pennies or, in very large companies, tens, hundreds, and even thousands of dollars. It is the number to the left that is most important. Even the U. S. government now permits each of us to figure our income tax in round numbers, to the nearest dollar for each deduction and part of the calculation. If your income-tax report is at all complicated and you do it yourself and have not tried rounding it off, you will be astonished next time you do it. It saves close to half the time of doing the report.

If any one technique in this entire book is worth more than the price of admission, I would be tempted to put the left-to-right methods of working first on the list. There are other valuable techniques, but the left-to-right methods are utterly unique.

The value of this approach to your number sense can only develop as you learn the methods that make it possible. The point to be made here is simply this: work at it. It is, as you learn to use it, as black-and-white a difference as thinking of the number 462 or approaching it as 2, 6, 4.

Convert to Simpler Forms

Most of us convert some of our figuring problems to simpler forms, when we can and when we notice that we can, without thinking very much about it.

You wouldn't give a second thought to wondering how much you had in terms of dollars if you found three 25¢ pieces in your hand. We call 25¢ a quarter because that is just what it is—a quarter of a dollar. In fact, if you take one out of your pocket right now you will find that it doesn't even say anything about cents. The official designation is "quarter dollar."

Whether anybody has ever called your attention to it or not, you are thinking now in terms of aliquots. An important chapter comes later on the short cuts that aliquots make possible. The whole concept, once you get used to it, is merely an extension and refinement of your instinctive understanding that 75¢ is the same as ¾ of a dollar.

This is conversion to a simpler form.

Perhaps, too, you have noticed that you can more easily multiply 692 by 99, by subtracting one 692 from a hundred 692's (69,200 − 692) than by setting up the whole problem with a pencil and paper and going through the classical form, which would look like this:

$$\begin{array}{r} 692 \\ 99 \\ \hline 6228 \\ 6228 \\ \hline 68508 \end{array}$$

Which is quicker and easier? Yet in doing the first you were merely using a basic and helpful form of the technique we call "round off and adjust." It can apply to many more numbers than 99.

This, too, is conversion to a simpler form.

Or perhaps, in quickly trying to come up with an appropriate tip for a meal check where 15% is standard, you noted that you could mentally take one-tenth of the check and then add one-half of that number to the one-tenth. A five-dollar check, for instance, would call for a 75¢ tip. One tenth of five dollars (50¢) plus one half of 50¢ (25¢), gives 75¢ quickly and easily.

It is obviously more convenient to arrive at 75¢ this way than to try (mentally or on the edge of the check) to multiply in the classic manner:

```
$ 5.0 0
    .1 5
--------
 2 5 0 0
 5 0 0
--------
.7 5 0 0
```

Yet in doing this little trick, you are merely practicing a fairly simple form of the short-cut method called "breakdown."

There are other useful forms of conversion, such as factoring and proportionate change. The application of these methods to number sense will become plain as you learn and begin to apply them.

The Four Steps to Number Sense

Here, for quick review, are the four steppingstones to number sense:

Practice seeing relationships
How does 5 relate to 10? 3 to 9?

See only the answer
Read 4 + 3 as 7—not as four plus three.

Work from left to right
27 is 27—not 7, 2.

Convert to simpler forms
25¢ is both 25¢ and a quarter of a dollar.
99 is 100 minus 1.
15 is 10 plus ½ of 10.
(And more conversions to come.)

Before going on to the first real "working" chapter of this book, get in practice for using it as well as reading it by trying to see only the answers to the following multiplications. Remember, 6 × 7 is 42—not six times seven:

2 × 9	7 × 3	8 × 4
5 × 5	4 × 6	3 × 2
4 × 7	5 × 8	3 × 9

2
COMPLEMENT ADDITION

IT HAS been estimated by experts that, for the average business, the total time spent in arithmetical computations breaks down to 70% addition, 5% subtraction, 20% multiplication, and 5% division.

These exact proportions may or may not hold in your particular business or profession. But chances are that they are not far wrong if you include all the number work you do.

So the obvious first job of becoming better at figures is to simplify by a very substantial margin that 70% of the time spent adding. What is simpler is, by nature, faster. Since adding is the single most-often-used process, it is worth spending a little extra effort at the beginning to learn a new approach that is guaranteed to make your work both easier and much, much speedier.

The approach you are about to learn is quite different from the one taught in any school. In fact, it has never even appeared in any of the books on the subject and is practically unknown in this country.

There is a reason for this. The reason is that the basis of this system is not part of our western civilization at all. The basis comes from Japan.

Back in 1946, an amusing story appeared in many American newspapers. The story said, incredibly, that in a contest in the Ernie Pyle Theatre in Tokyo the most expert electric calculator operator of General MacArthur's headquarters had been roundly defeated in a public match by—of all things—an abacus!

In a long series of problems, ranging from addition and subtraction of as many as fifty numbers with three to six digits each, through division and multiplication problems with up to twelve digits each, the electric calculator had gone down to resounding defeat. The winner was a "primitive" instrument of beads on rods.

An abacus is really nothing more than a recording, not a calculating, device. It is basically so simple and useful a machine that different forms of it were used in Rome, India, China, Japan, and many other countries. The varieties used have been very different indeed, some of them about as clumsy as they were useful, but in Japan the highest mathematical thinking was brought to bear on the problem. An entirely new, "streamlined" version called the *soroban* was developed within the last few decades.

The soroban still consists of beads on rods. This is basic to anything that can be called an abacus. But it has fewer beads on each rod than any other variety. Where some contemporary Chinese models still have as many as fifteen beads on each rod, the soroban has exactly nine.

The number nine rings a bell. It is the highest of all single-digit numbers . . . the basis of our decimal (tens) counting system.

The Japanese mathematicians saw this fact. After thousands of years of using the device in their calculating, they sat down and realized that it was silly to record ten or more on any one rod, because that ten could be recorded on another rod with just one bead in precisely the same way that we record a ten on paper—with a one moved over one place to the left.

Actually, of course, the electric calculator in that Tokyo contest was not defeated by the abacus at all. The operator of the calculator was defeated by the operator of the abacus—a man trained in the Japanese system of soroban arithmetic, which is so much simpler and faster than ours that he could solve and record each step of a problem faster than the electric calculator operator could punch them into his keyboard.

The soroban operator was no number genius, incidentally. He was a champion operator, but (as he himself stated) no better than many other first-class operators. After all, the soroban is still the basic tool of Japanese arithmetic, which today is building an industrial complex producing the most sophisticated binoculars and cameras and advanced radios.

If today you want a number job in Japan, don't bother to learn how to operate an adding machine. Learn the soroban.

Soroban Theory

The soroban, or modern Japanese abacus, is useful to us here because it is a valuable tool for calculating in its own right and because in order to use it with such incredible efficiency and speed the Japanese had to develop the theory.

Three parts of this theory are especially useful and applicable to our technique of streamlined arithmetic:

1. Do each step one at a time, recording the results in the quickest and easiest way.
2. Work from left to right.
3. Never calculate over ten.

That last one is a surprise. It surprised me some years ago when I was researching the whole field of short-cut mathematics for a program I was editing and, remembering that story about the Tokyo contest, I did some research on modern soroban theory.

Never add over ten? The whole idea violates everything we learned in school and everything we think we know about numbers. At first sight, the method for doing so will look more complicated. We are tempted to dismiss the idea and go on to something else.

But it does make sense. It makes enough sense for a soroban operator to beat the pants off an electric calculator operator.

Never add over ten. It takes time to get used to this idea.

If you react as I did when I first read the theory and method, then applied it to streamlined math and found how well it worked, you will need several days to adjust to the concept. But use it anyway. Force yourself. At first it will take longer than the way you now do arithmetic, because you will be breaking old habits and building new ones: new ones you'll prize for the rest of your life. Soon, if you keep working at it, you will find that you can do problems far more quickly and accurately than you have ever done them before.

Never add over ten! What about 5 + 6? 8 + 3? 9 + 7? We will get to that very shortly. Before going into it, though, you should understand thoroughly why this system is so fast.

Even though you have already memorized the addition tables up to 9 + 9 or even more, you will gain tremendously if from now on you concentrate on just about half of them—the easier half, at that. Soon you will naturally, almost unavoidably, become almost twice as fast on the easier half you really use.

Combine this with an automatic-recording system for taking care of the tens, such as the soroban provides or the two techniques developed especially for this system, and your speed accelerates still further.

Look at the following table of all possible combinations of two digits. You will find that there are forty-five of them in all, from 1 + 1 to 9 + 9. Now notice that of the forty-five combinations, twenty add up to less than ten. Five add up to ten. Twenty add up to more than ten.

The twenty combinations that add up to more than ten, incidentally, are also the twenty hardest to remember quickly and the ones on which most of us stumble most often.

The table, incidentally, shows each pair only once. That is, 2 + 5 is shown in the "two" column but 5 + 2 is not shown at all; it is merely the same pair backwards.

POSSIBLE DIGIT COMBINATIONS									TOTAL OF EACH PAIR
1 + 1									2
1 + 2				less than ten					3
1 + 3	2 + 2								4
1 + 4	2 + 3								5
1 + 5	2 + 4	3 + 3							6
1 + 6	2 + 5	3 + 4							7
1 + 7	2 + 6	3 + 5	4 + 4						8
1 + 8	2 + 7	3 + 6	4 + 5						9
1 + 9	2 + 8	3 + 7	4 + 6	5 + 5					10
	2 + 9	3 + 8	4 + 7	5 + 6					11
		3 + 9	4 + 8	5 + 7	6 + 6				12
			4 + 9	5 + 8	6 + 7				13
				5 + 9	6 + 8	7 + 7			14
more than ten					6 + 9	7 + 8			15
						7 + 9	8 + 8		16
							8 + 9		17
								9 + 9	18

The appearance of this table is not random. It could be set up in slightly different shapes, but the order and pattern of this particular arrangement are especially instructive. You will find it worthwhile to examine the pattern with some care. Note, among other things, the heavy concentration of pairs adding up to totals around ten, and how the possibilities taper off toward high and low totals.

In the system about to be explained, here is how we will handle the forty-five different combinations:

We use the twenty combinations adding up to less than ten just as we do now. They are the easiest ones. We use the five combinations that add up to ten (1 + 9, 2 + 8, 3 + 7, 4 + 6, and 5 + 5) even more than we do now, so we learn them extra well. We *forget* those twenty hardest combinations that add up to more than ten and learn the technique of complement addition.

Add With Complements

The basic rule for the new technique is this:
To "add" over ten, subtract the complement of the larger digit from the smaller digit—and record a ten.

First we need to learn what complements are. Then we will take up how to record tens. Both are easy.

What is a complement? A complement is simply the digit that, added to the digit you have, adds up to ten. You might say that a complement is that digit needed (in addition to the one you have) to *complete* a ten.

For example: The complement of 9 is 1, because 9 + 1 is ten. The complement of 8 is 2, because 8 + 2 is ten. The complement of 7 is 3, and the complement of 6 is 4, because 7 + 3 is ten and 6 + 4 is ten. Even in your sleep you would answer that the complement of 5 is 5.

Those are all the complements you ever have to remember in adding the longest column of figures. There are only five of them: five pairs, you will note, that add up to ten in the table of possible combinations.

Before learning how to add with complements, make doubly sure that you have the idea by looking at the following digits and giving their complements. Try to "read" the complement of each as you are beginning to "read" the answer to a simple problem:

7 6 9 5 8

The way you add with complements takes a bit of getting used to. But it is one of the most fascinating and fruitful approaches known to short-cut arithmetic. You "add" two digits that total more than ten by subtracting the complement of the larger digit from the smaller digit and recording a ten.

In order to add 6 + 7, you subtract the complement of 7 (3) from 6, and record a ten. 6 − 3 gives 3. The recorded ten makes it 13.

Or to add 8 + 4, you subtract the complement of 8 (2) from 4 and record a ten. 4 − 2 gives 2. The recorded ten makes it 12.

It is useful to subtract the complement of the larger digit rather than the complement of the smaller. In this way you cut in half the number of complements you have to remember at this stage—though the other half of the complements are really only the same pairs of digits that add up to ten turned around. Just as 2 is the complement of 8, so is 8 the complement of 2.

Try it yourself, before going any further. Add 7 + 9 by subtracting the complement of the larger digit from the smaller digit. The complement of 9 is __. 7− __ is __. Remember to record a ten, in ways you will learn very soon. So the answer is 16. I hope that is what you arrived at through the new method, even the first time. If not, then it hasn't become clear yet. Another reading of the last few pages is indicated.

Now add 3 + 8. Would you subtract the complement of 8 from 3? What is the complement of 8? Don't forget to record a ten.

Strange and complex as this system undoubtedly seems at the moment, it is really far faster. This is because you are working with only the easier half of the forty-five digit combinations, the half that add up to less than ten. Even subtracting the complement will shortly become no problem, because you are always subtracting digits from pairs in the top part of the table. Look back at it again for a moment. In the complement system of addition, you *cannot possibly* get into that bottom part of the table—those twenty toughest (and slowest) combinations.

Give it one more try before going on. Each time you use it, the system will become a little easier and more natural.

Add 6 + 5. The complement of 6 is __. 5 minus __ is __. Record a ten.

Recording Tens

You recall that we said the soroban, or modern Japanese abacus, is not really a calculating instrument at all. It is a recording instrument. By recording the results of each step in a calculation, it frees the operator to concentrate on skill and speed in taking each step.

We can record steps, too. Our methods of recording will enable us to concentrate on speed, just as the soroban does.

There are two good ways to record tens each time you use complements. The first way is simply to put a line at each place in a column of figures whenever you use a complement or add to ten. If you adopt this system, make it a habit so it becomes automatic. Then, when you write your final total, you just sweep your eye over the lines in that column and put down the total number of lines as your "tens" digit, one place to the left. Instead of remembering "37," for instance, you have in your mind at that point only the *single digit seven,* but you will find three lines along the column.

We will go through one problem slowly and carefully, step by step. At first, the process will seem quite long and complicated because each step must be made clear. Actually, as you will find with use, it is far simpler and faster than the traditional method of addition.

Remember that we always work from *left to right:*

```
  2 4
  3 5
  9 9
  7 9
2 1 7
  2
-----
2 3 7
```

First column

2 + 3 is 5.

5 − 1 (complement of 9) is 4. Record ten by putting a line under the 9.

4 − 3 (complement of 7) is 1. Record ten by putting a line under the 7.

Put down the remembered 1 under the column.

Count the lines. There are two. Put a 2 one place to the left of the column.

Second column

4 + 5 is 9.

9 − 1 (complement of 9) is 8. Record ten by putting a line under the 9.

8 − 1 (complement of 9) is 7. Record ten by putting a line under the 9.

Put down the remembered 7 under the column.

Count the lines. There are two. Put a 2 one place to the left of the column, under the 1 from the first column.

Now you simply add and get the answer, 237.

While this has taken some time to explain step by step, in practice you will find it infinitely faster than the old way. When you do it automatically, you will think only "5, 4 (line), 3 (line); 9, 8 (line), 7 (line); 237."

One element about the problem may be a little confusing. We combine the next figure in the column with the figure in our mind from previous additions, not with the figure above it. For instance, in the first column of the problem above, we subtract the complement of 9 (1) from 5—the result of adding 2 and 3—not from the 3. It works just like regular addition in this respect. The use of complements does not change it.

Try the next example, in which we will go through the steps in a much more condensed way. See if you can follow each step, identify the complement being used in each case, and understand why we record a ten with a line each time we do so:

$$\begin{array}{r} 6 \\ 2 \\ \underline{8} \\ \underline{7} \\ \hline \end{array} \qquad \begin{array}{c} \text{8, 6 (line), 3 (line).} \\ \text{Answer: 23} \end{array}$$

This example should have gone a little more easily. Take it slowly now, so you can build on a solid base of thorough understanding in later parts of the book.

Rather than go on with more practice at this point, let us get into the second method of recording tens. Of the two, this is quicker and more generally useful. But, in this case and in many alternate choices in the "short cuts" section later in the book, you should adopt the one that seems most natural to you and concentrate on it. Continuous use of one system will build the desirable habit pattern and accelerate your speed.

Record on Your Fingers

The second way to record tens is to use your fingers. We were taught not to count on our fingers, so the idea may come as something of a shock. Actually, however, the purpose here is vastly different. We were taught not to count on our fingers because using them for *counting* is leaning on a crutch that interferes with genuine mastery of the calculating skill itself. Using them for *recording,* as you will see, approaches the automatic-recording advantages of the soroban, and frees you to concentrate on adding the digits with extra speed.

Should you need any more encouragement, take note of the fact that top abacus operators become amazingly proficient at mental arithmetic by learning to close their eyes and visualize the soroban as they calculate—and they use their fingers for recording. So no matter how much distaste for using your fingers your school training may have left you, keep firmly in mind that this is recording rather than counting, and give it a try. Speed mathematics can and should make use of any device that simplifies and speeds up the solving of problems.

Here is how the system works. To record the first ten (when you first use a complement or add to ten), fold the little finger of your left hand into the palm. If you write with your left hand, there is no reason why you cannot record on the right. To record the second ten, fold the next finger alongside the little finger. This means two tens. If you use another complement or add to ten in the same column, fold the next finger. This records three tens. And so on, up to five tens.

If you have more than five tens in a long column, open the hand and start over with the little finger again. Perhaps you will feel happier about remembering to add five to the second running total of tens if you make a line in the column when you start over. Or use any other signal to yourself that makes sense.

This is not silly. Any mechanical aid that fits your habits and personality is a valid and useful device for freeing your mind to concentrate on the basic objective: speed and ease with fingers.

Whatever signal you adopt in a case like this, be consistent with it. Settle down to use this method for every single calculation you do, no matter how simple it is or where you do it. Habits are very important. Making a habit of consistently using the fastest techniques is what gives speed.

The use of fingers instead of lines to record tens does not

change what you do at the end of each column, of course. First you put down the digit in your mind from the final addition. Then you put, one place to the left, the number of fingers you have folded—adding five if you had to start over again.

Here is how we solve a problem with this system. Work from left to right:

First column

5 + 1 is 6.
6 + 3 is 9.
7 − 1 (complement of 9) is 6. Fold a finger.
Put down the 6 in your mind.
One finger folded. Put down 1 one place to the left.

```
  5 7 4
  1 9 8
  3 6 5
  7 4 4
  -----
1 6 6 1
  2 2
-------
1 8 8 1
```

Second column

7 − 1 (complement of 9) is 6. Fold a finger.
6 − 4 (complement of 6) is 2. Fold a finger.
2 + 4 is 6.
Put down the 6 in your mind.
Two fingers folded. Put down 2 one place to the left, under the 6 from the first column.

Third column

4 − 2 (complement of 8) is 2. Fold a finger.
2 + 5 is 7.
4 − 3 (complement of 7) is 1. Fold a finger.
Put down the 1 in your mind.
Two fingers folded. Put down 2 one place to the left, under the 6 from the second column.

Note especially that, because it is a faster habit to use the complement of the larger of the two digits to be added at any point (one being in your mind from the last addition, the other being the next digit in the column), sometimes you use the complement of the digit in your mind, and sometimes the complement of the next digit in the column. It makes no difference.

Now we will go through another example with a condensed explanation of the process:

```
    7 8 2
    4 2 6
    9 8 4
    3 6 1
    5 7 9
    -----
  2 8 1 2
    3 2
  -------
  3 1 3 2
```

First column

1 (finger), 0 (finger), 3, 8. 8 under the column, 2 one place to the left.

Second column

0 (finger), 8, 4 (finger), 1 (finger). 1 under the column, 3 one place to the left (below the 8).

Third column

8, 2 (finger), 3, 2 (finger). 2 under the column, 2 one place to the left (below the 1).

This example demonstrates one new fact. In developing the final answer you sometimes have to raise a digit you have already put down. In the problem above, the 2 in the very left column becomes 3 in the final answer of 3132. Since you are adding just two lines at this point, it should not be a problem. When we get into multiplication, where it can be a little harder, you will learn a special recording technique that makes it possible to work from left to right with quite complex problems in this way. But in adding you never have to add more than two lines, and no digit in the final answer ever needs to be raised in value by more than one. You should be able to work from left to right by merely glancing at the next column as you put down each digit to see if the total of the next column will be ten or more. If it will be (you don't care how much more than ten it will be at this point), just add one to the digit you are about to put down.

In the problem above, you glance at the second column and note that 8 + 3 will be more than ten. So instead of putting down 2 as the first digit, you put down 3. In a sense you are pre-recording a ten from the complement you will use when you get to the second column. For the second digit of the final answer, you subtract the complement of 8 (2) from 3 and put down 1. The ten has already been recorded by raising the first 2 to 3.

Why Complements Work

The use of complements is at the very heart and center of modern abacus theory in Japan, where today the soroban rather than the adding machine stands on the average bookkeeper's desk.

You don't have to understand the theory of complement addition to use it, but understanding always helps mastery. Learning simply by rote leads to a shaky mastery at best—to what W. W. Sawyer calls "imitation" instead of substance. So let us take apart the theory of complements and see why they work the way they do.

Since our counting base is ten, any addition is really a process of going up to ten and then *starting over again*—recording a ten by remembering "xxteen," "twenty-xx," and so on; or, with our new system, by using a line or a folded finger.

When we add two digits that would go over ten in complement addition, we really do just what a soroban operator does when he has to add some beads to a rod and finds that there are not enough beads on the rod. The streamlined abacus, or soroban, has only five beads on a rod: one representing a value of five, and four each representing a value of one. Altogether, they can record no more than nine.

Suppose the operator has recorded eight on one rod. Beads are moved toward the center divider in order to record, and a total of eight on one rod would look like this:

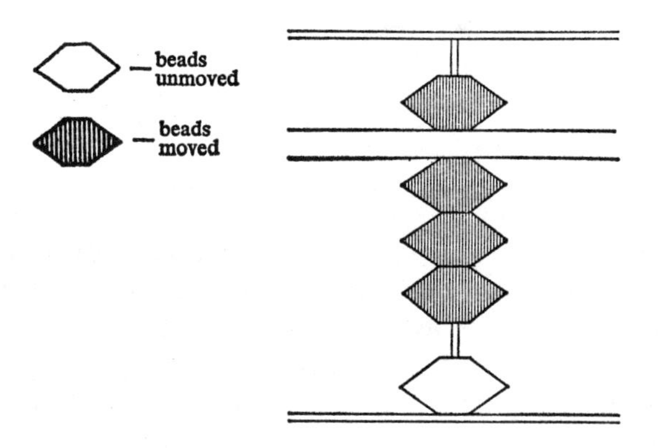

The five-bead is the one above the separator. It is moved to the center in order to record a five. Three one-beads have been moved toward the separator. This rod is recording the number eight—five plus three.

Now suppose the operator has to add nine to this number.

He can't. There is only one bead not recording (the one on the bottom) and that would add only one. How can he add nine?

This is where modern abacus theory took over in Japan. Mathematicians developed the approach that the operator should never *try* to add more beads than he can find on the rod—even in his head, which was the way it had been done before. Instead, he should *subtract* the *complement* of the new digit, and record a ten on the rod to the left.

So, in order to add nine to the eight recorded above, the operator—knowing his complements cold, as he must—merely flicks one bead *away* from the separator and immediately flicks one bead on the rod to the left toward the separator to record the ten.

After he subtracts one (complement of nine) from this rod and adds one (ten) on the rod to the left, the answer looks like this:

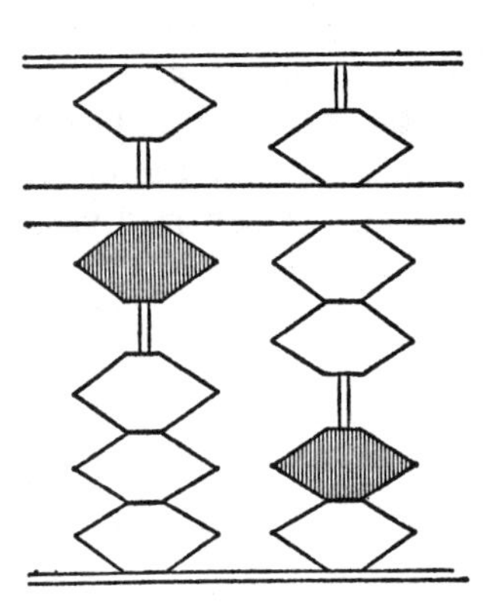

Simple? Yes. But very subtle, and very revolutionary to our ways of doing arithmetic. The answer on these two rods is 17; one ten plus one five plus two ones. But it was produced *without* ever adding eight plus nine. It was produced by subtracting the complement of nine (one) from eight and recording a ten.

Soroban teaching calls this "letting the answer form naturally on the board." What we are learning to do, in our mental adaptation of soroban theory, is let the answer form naturally in our mind.

Let us go a little more deeply into the theory of complements, in order to reinforce still further your "number sense" in using them.

Remember that each time we add beyond ten we start over again with one—11, 21, etc. Since using an addition table reaching beyond the next ten only compounds the number of possible combinations we must memorize and handle with ease, the use of complements enables us to deal only in combinations of ten or less and yet run through the entire counting system.

Take an example for which we would not normally use the complement system. You can add ten and nine in either of two ways:

$$\begin{array}{r} 1\ 0 \\ +\ 9 \\ \hline 1\ 9 \end{array} \qquad \begin{array}{r} 1\ 0 \\ +\ 1\ 0 \\ \hline 2\ 0 \\ -\ 1 \\ \hline 1\ 9 \end{array}$$

This is very easy to understand at sight. 9 is 1 less than ten, so we can just as well add ten and subtract 1 as add 9. This is true no matter to what other digit we add it:

$$\begin{array}{r} 5 \\ +\ 9 \\ \hline 1\ 4 \end{array} \text{ or } \begin{array}{r} 5 \\ +\ 1\ 0 \\ \hline 1\ 5 \\ -\ 1 \\ \hline 1\ 4 \end{array} \qquad \text{as well as} \qquad \begin{array}{r} 8 \\ +\ 9 \\ \hline 1\ 7 \end{array} \text{ or } \begin{array}{r} 8 \\ +\ 1\ 0 \\ \hline 1\ 8 \\ -\ 1 \\ \hline 1\ 7 \end{array}$$

This works because, as you already know, 1 is the complement of 9. Working out each step of the theory, the complement approach may appear more complicated. Working out the addition of ten is what makes it appear to be so; we never bother to add ten as such, because we can simply record it.

Done in this fashion, the two above examples now look like this:

```
  5          5          5
+ 9  or  + 1 0   or   - 1
---      -----        ---
1 4        1 5          4 and record
         -   1            a ten, which
         -----            makes it 14
           1 4

  8          8          8
+ 9  or  + 1 0   or   - 1
---      -----        ---
1 7        1 8          7 and record
         -   1            a ten, which
         -----            makes it 17
           1 7
```

Can you *feel* the identity of all three processes?

We chose 9 as the demonstration example because it is so obviously 1 less than ten. Just as surely as 9 is 1 less than ten, 8 is 2 less than ten, 7 is 3 less than ten, and so on. The principle does not change one bit when we use these other combinations.

As one further example, let us show all three ways of expressing another "identity":

```
  4          4          4
+ 7  or  + 1 0   or   - 3
---      -----        ---
1 1        1 4          1 and record
         -   3            a ten, which
         -----            makes it 11
           1 1
```

Take your pad at this point and work out the addition of 5 and 8 in all three ways. The closer you can come to "feeling" the identity of all three pictures of the same process, the more confidently you will handle complements.

One analogy that has proved helpful to some people is to visualize the process of adding as climbing a series of ladders, each with ten rungs, from level to level. At any point, you know your position on a ladder and you know on which ladder you stand. For instance, you are now standing on the sixth rung of the third ladder—an analogy of the number 36. You are told that you can advance eight more rungs, and wish the quickest and easiest way of projecting where you will be standing after eight rungs.

First, you know that you will be on the next higher ladder (in the 40's), because there are not eight more rungs above you on the third ladder. Adding 6 and 8, let us say, is something you have never been taught to do. You do know that if you could advance a full ten rungs you would be on the corresponding rung of the fourth ladder—46. But since 8 fails by 2 to complete a ten, you will be 2 rungs lower—44.

So, in any addition that crosses the next ten-point, you will fail to reach the corresponding number across that ten-point by precisely the amount that the number you add fails to reach ten. That is its complement.

Before going on to the next chapter, work for a few moments at making the use of complements a habit by using them conscientiously in adding the following problems. Use either lines or fingers as you prefer, but standardize now on one system or the other.

Do not add these pairs. In each case, subtract the complement of the larger digit from the smaller *and* record a ten. Just "see" the answer; don't write it down:

```
5    8    4    3
7    6    9    8
—    —    —    —

9    4    6    7
2    7    5    9
—    —    —    —
```

As you "read" these examples (and you should be trying to "read" rather than "solve" them) it may help to channel your thoughts in the right direction if you lip-read them the first few times. This is not good permanent practice, but it will help break your old habit patterns. You would lip-read the first problem, for example, as "5 − 3 is 2; finger," to help you avoid slipping back into the thought pattern of "5 + 7 is 12." Ultimately, you will try to "see" it as merely "2, finger."

The first key to speed in this system is obviously knowing your complements at sight, without pausing to think for a second. Review them quickly. Try to "read" the complement of each digit as you see it, without stopping to ponder:

8 7 9 6

These are the only digits for which you have to remember complements at this point. Five is the complement of five, but you never use it that way in adding because when faced with a five and a five you simply react "0, finger." When faced with a five and a larger digit, you use the complement of the larger digit.

What is the complement of 7?

If you had to pause for even a flicker, build your base for rapid progress later in the book by reading the above digits again. React without thought with the complements to these digits:

6 9 8 7

The sheer repetition here is not overdone. It is essential to mastering the new system. One of the two major approaches to teaching machines uses precisely this principle.

Go through this brief check-up to make sure you are ready for the next chapter, which will begin to build your speed and confidence in complement addition.

What is a complement?

What is the complement of 7?

When you add two digits that would go over ten, do you add or subtract the complement of one of them?

Is it quicker to use the complement of the larger digit, or the smaller one? Why?

What is the complement of 6?

When adding a column, do you keep all the tens in your mind, or do you record them?

What is the complement of 8?

How do you record a ten?

What is the complement of 9?

In adding a column, do you combine each new digit with the digit above it, or with the digit in your mind from above additions?

Could you explain to a friend why complements work as they do? Pretend he has just asked you, and see if you can.

3
BUILDING SPEED IN ADDITION

IN THE last chapter you have had a taste of one of the newest and most exciting developments in the whole field of speed mathematics. Its sheer beauty and rapidity will grow on you as you begin to make it a habit.

Part of making it a habit is plain old-fashioned practice. There is simply no way of learning high-speed arithmetic without a pretty fair dose of practice. You cannot begin to master the systems without using them enough times to feel at ease with them.

It is always a temptation to skip the practice in a book of this kind. You are interested in the "meat," in the theories, in what comes next. There is a great deal coming next. But to skip the practice in its proper place would be unfair to yourself. The best theory, the finest technique in the world, is useless unless you can use it. You cannot use it simply by knowing the theory. The difference between knowing how something is done and knowing how to do it is skill. Only practice can build skill.

We will vary the practice, break it up into modest doses, to keep it as inviting as we can. But—don't skip it!

In order to encourage you to do the practice page by page, I have hidden right in the middle of it one more big step for even greater speed in addition.

Start now by reading at sight the answers to the following additions. Don't think or lip-read or even "see" the problem itself if possible; see only the answer. Remember your complements for groups that would go over ten:

7	4	1	6	8	4
4	4	2	4	5	3
—	—	—	—	—	—
3	8	2	9	1	2
6	2	5	3	4	7
—	—	—	—	—	—
5	9	1	5	6	4
3	5	8	8	6	1
—	—	—	—	—	—
6	3	8	9	5	7
7	2	9	7	5	2
—	—	—	—	—	—

Pause and ask yourself some questions here. Did you manage to see only the answer, not the two digits to be added? Did you begin to find yourself glancing at each group that would add over ten and automatically subtracting the complement of the larger digit from the smaller digit—and folding a finger?

If not, go back over them and make the special effort to use complements in these cases. Such combinations are mixed in with "under ten" combinations on purpose. The two are always mixed in the figure work we meet in our lives.

Now let us go on to another easy dose of practice. These numbers are not simply random, by the way. Every possible combination of digits has been recorded and appears in the practice tables. By the time you finish this chapter you will have practiced every single possibility.

See only the answers to these, using complements where appropriate:

8	7	4	3	6	4
7	3	9	3	1	6
—	—	—	—	—	—

3 1 6 5 3 2
4 1 5 9 5 8

2 7 1 8 8 6
9 8 6 4 1 9

5 4 3 5 2 2
2 5 8 7 2 1

That's enough for a moment. Arithmetic, even the streamlined variety, takes concentration. At the start, the new techniques take even more concentration than the old ones, because you have to stop and think about doing things in the new way.

Before finishing the random series of all digit combinations, take a breather by hearing the famous (and possibly apochryphal) end to the story of that Tokyo contest between the abacus and the calculating machine. The electric calculator, according to the story, was made by International Business Machines, whose company-wide motto is THINK.

After the American machine-operator was roundly defeated by the soroban-operator, he is reported to have said: "Maybe his way is faster. But all I have to do is punch buttons. He has to think."

Now we will finish up our speed practice in basic digit combinations. Remember to use complements where the addition would go over ten, and fold a finger or think "line":

1 9 3 5 1 3
9 2 9 6 7 1

4 9 8 9 6 3
2 8 6 9 8 7

6 5 4 8 1 4
3 4 4 3 3 8

7 2 7 6 9 5
6 3 5 2 1 1

7	2	9	8	9	2
1	4	4	8	6	6

1	7	7	9	7	4
5	7	8	8	9	9

That's all. Those are all the possible digit combinations. You will never, in all your life, face any combination of digits that you haven't just practiced.

Some of the addition we do in our jobs or at home consists of single pairs, such as the examples you have just done. Much of it, however, does not. We frequently have to add three or more digits in each column of a particular addition, whether it is sales in seven different territories or prices of twelve lots from the real-estate developer.

Handling more than two digits using the complement system is something you already understand but might profitably use a little practice on. This involves handling complements when one of the digits to be combined is in your mind (from adding the previous digits in the column) and the other is the next digit in the column, rather than with two digits set up just for you to practice with.

Consider this addition:

$$\begin{array}{r} 7 \\ 7 \\ 7 \\ \hline 21 \end{array}$$

When you add the first two digits, you subtract 3 (the complement of 7) to get 4, and record a ten—14. The only digit you carry in your mind, however, is 4. The ten you record with a line or a folded finger, and promptly ignore for greater speed and accuracy.

Now you glance at the last 7. You combine it, of course, not with the 7 above it but with the 4 in your mind. $4 - 3$ (complement of 7) is 1, with another recorded ten. You have recorded two tens and are remembering 1, so your answer is 21.

This answer "formed itself naturally" in your mind, just as it forms itself naturally on the board of the soroban.

While you know all this, you will handle the process more easily and quickly if you spend a few minutes consciously practicing the use of it. Run through the next column with the complement technique. Then see if your handling agrees with the description below it.

$$\begin{array}{r} 9 \\ 8 \\ 6 \\ 8 \\ 9 \\ 5 \\ 7 \\ \hline 5\ 2 \end{array}$$

The complement system, assuming you use fingers (if you use lines, read "lines" for "fingers"), would go like this: "7 (finger), 3 (finger), 1 (finger), 0 (finger), 5, 2 (finger). 5 fingers plus 2—52."

Note especially that between the 5 and the last 7 there is no finger. Why?

Now read through these examples, using complements in each case and seeing if the total of your recorded tens plus the number in your mind comes out the same as the answer. If not, do them again:

$$\begin{array}{rrrr} 7 & 3 & 8 & 4 \\ 5 & 9 & 7 & 2 \\ 8 & 9 & 6 & 0 \\ 9 & 9 & 9 & 2 \\ 3 & 9 & 6 & 1 \\ \hline 3\ 2 & 3\ 9 & 3\ 6 & 9 \end{array}$$

The last one was put in there on purpose, just to remind everyone that we don't *always* use complements. They only apply when addition goes over ten.

Compound Your Speed by Grouping

You have learned, and begun to practice, two basic elements of real addition speed: recording tens, and using complements instead of adding over ten.

There is one other major contributor to high-speed adding. It is a standard "short-cut" method. But it is easier than ever to use with complement addition, because you will get to know the twenty-five combinations to which it most easily applies by first name, instead of scattering your memory over all forty-five possible combinations.

Your full mastery of those twenty-five easiest combinations can speed up your addition still further if you stretch it to include the technique called grouping. In grouping, you "see" any pair of digits adding to less than ten as one digit, and any complementary pair as leaving the number in your mind unchanged but worth another recorded ten.

Just as you look at the two letters *i* and *t* and see—not *i* and *t*—but "it," so you will learn to look at "3" and "4" and (if you are adding) see only "7." It works like this:

3, 4	} 7	An expert will handle this as the addition of 7, 8, (record), and 8. He will "see" the 3 and 4 as 7, and so on. Simply think "5 (finger), (finger), 3 (finger)—33." It's fast—and surprisingly easy.
2, 6	} 8	
3, 7	} – (record)	
5, 3	} 8	

Now try grouping on these examples:

2	7
7	1
6	8
3	2
1	3
4	2
5	6
2	3
7	6
1	3
3 8	4 1

In any future addition examples, make a special effort to group digits that add up to less than ten as well as to ten exactly. Steady work with complements will help flag 3 plus 7 as worth exactly one folded finger (or one line), without changing the number in your mind from previous additions.

All your adding practice so far has been single-column work. Some of the adding we do in our jobs or at home is of this nature, but it is more than likely that a large part of it includes several columns.

Now is the time to refresh your memory on working from left to right. The abacus is always used this way. That Japanese operator who so thoroughly beat the calculator operator would not dream of working from right to left. It just would not be natural.

Remember that when we add several columns, we put down under each column the last digit that developed naturally in our mind, and one place to the left of it we put the number of recorded tens. Under the first column we can place our recorded tens immediately to the left, but under later columns they have to go down one line because of the totals of those columns. Follow, using all your new techniques, this example and see if your answer agrees. Work from left to right:

```
  8 9 , 5 2 8 , 8 1 7
  9 7 , 7 7 6 , 9 4 2
  9 4 , 8 7 6 , 2 4 9
---------------------
2 6 0 , 0 6 0 , 9 9 8
  2 2   1 2 1     1
---------------------
2 8 2 , 1 8 2 , 0 0 8
```

This example shows one or two special points. Note that in the next-to-last column, there are no tens recorded and therefore there is no digit placed to the left of that column. Note also that in adding the two sub-totals, you carry one "ten" back from the next-to-last column, through the column before that, to the column before that one. When you come to adding your sub-total lines, you will sometimes have to do this. Since you never add more than two lines of sub-totals, a glance ahead will show when you need to "carry back" a ten. If this proves difficult, simply *underline* a digit to which you find you have to carry back a ten. The underline raises the value of the underlined digit by one—a technique you will learn to use automatically when we get to multiplication.

Using this method, the final answer to the example above would look like this:

$$2\ 8\ 2\ 1\ 8\ \underline{1}\ 0\ 0\ 8$$

You underline the 1 because you have looked at the next column before putting it down and seen nothing to carry back. But when you add that next column (the 9 with nothing under it), you see that you will have to add a ten from the next-to-last column—the 9 plus 11—and this will change the 9 to a 0, with a ten carried back to the 1 you have already put down. It would be awkward to change the 1 by this time, so you simply underline it. In reading or copying the final answer, read the $\underline{1}$ as 2.

If this seems hard or slow, note that the same thing often happens when you add or multiply on the abacus; and it is considered more than worthwhile to carry back a ten in this fashion rather than pay the far greater price of working from right to left.

The obvious job remaining is to practice a bit more; practice so that the techniques become second nature, so that you begin to "see" only the answer, so that you group digits adding to ten or less without having to think about it.

Try reading right through the following problems, using all your newly learned techniques and noting your answers on your pad or cards for later reference:

1 5 8 4	8 6 7 1	2 6 3 8
7 3 1 3	9 9 6 5	5 7 4 4
1 9 3 1	6 8 9 3	6 7 3 6

At this point your practice is beginning to combine all the separate elements you have learned. Some columns involve complements and recorded tens; some do not. Some columns require you to carry tens back to a previous column in the final answer; some do not. Some columns contain digits you can combine at a glance; some do not. This is the variety of which our daily arithmetic is composed. It never comes in neat parcels designed especially to illustrate some special point.

Now go back, with a fresh page of your pad, and do the examples over again.

Compare the answers you got the two different times. Are they the same, or different? If you have two different answers in any case, do it still once again—and find out where you went wrong.

Now go on to these:

```
1 2 9      6 7 5      9 4 5
6 8 2      4 8 9      4 8 7
6 4 2      5 3 3      1 1 3
2 1 8      8 3 6      7 2 5
5 3 1      1 5 6      9 2 4
-----      -----      -----
```

Note your answers as you did before. These examples have fewer columns but more digits in each column. The variety is planned, in order to show examples of different applications of the techniques and to keep the practice from becoming too monotonous.

Now turn your pad or card over and do the above problems again. Compare your answers to the ones you got the first time around. If they are the same, good. If not, learn from your mistakes by doing any problems to which you got different answers once more, and seeing which one is really right.

Because it is so important to everything you will do for the rest of your life in mathematics, review right now the twenty combinations of digits under ten. Other than complements, they are the only ones you have to handle from now on. Combine these pairs *at a glance*:

```
1    4    2    8    3
5    4    6    1    2
-    -    -    -    -

1    3    2    4    1
4    5    1    3    1
-    -    -    -    -
```

7	3	1	2	3
2	3	6	4	1
5	4	1	3	2
2	5	7	6	2

This table includes *every* possible digit combination in adding other than complement pairs. The complementary pairs, too, should be starting to feel as natural as breathing. Look at the following digits and, in a flash, see only the complement:

7 6 9 8

As a finale to this chapter, try your hand at one really big problem—the sort most of us approach with some reluctance when we have to solve it, yet which combines in just one practice session everything you have learned so far. Approach it with these rules in mind: first, work from left to right; second, add "over" ten by using complements and recording the ten; third, record the tens as you go; fourth, combine digit-pairs adding to ten or less at a glance and handle them as a single digit or recorded ten.

Work for speed on this one. Note down your answer, and come back from time to time to see if on another try you still get the same answer. Vary your practice by adding down one time, adding up the next:

741,684,382
193,513,498
912,591,564
693,654,814
358,861,638
727,695,635
957,729,252
2,117,291
87,436
68,692
431

Do this at least once before going on. It embodies, in one example, every possible technique from the last two chapters.

4

COMPLEMENT SUBTRACTION

SUBTRACTION is merely the other side of the coin of addition.

For most of us, however, it causes far more trouble. There are probably two reasons for this. While many of us learned our "addition tables" by heart in school, few of us really mastered the conversion of these into "subtraction tables" with anything approaching the same thoroughness. More important, however, the traditional process of "borrowing" is a tricky concept. Many of us find ourselves forgetting to borrow, or borrowing twice, because it is basically unnatural.

This chapter will eliminate both these handicaps. It brings to your work in subtracting three important aids to speed and accuracy.

First, complement subtraction will enable you to work from left to right. This is quite impossible in any other method of speed mathematics, but, surprisingly, the left-to-right procedure works *best* with complements. You should begin to have some feeling at this point of how much left-to-right working helps preserve and build your number sense.

Second, you will use a new technique that does away with "borrowing" entirely. The same necessary step will develop naturally and easily in your answer, just as it does on the abacus.

Third, you will apply to subtraction the same complement technique you have just learned for addition. This means that never again will you have to subtract a larger digit from a smaller—the process that causes so much confusion and error. Just as you now do in adding, you will work entirely with the twenty easiest combinations and the five pairs that "complete" tens—and forget the twenty hardest combinations altogether.

Before getting into the complement portion of subtraction, it will be helpful to get used to handling subtraction from left to right on a few problems in which you can work from left to right with standard methods. Such problems are those in which each digit in the smaller number is smaller—or the same size as, but never larger—than its corresponding digit in the larger number. In other words, in any problem that does not involve "borrowing" you can as easily work from left to right as from right to left:

```
  9 7 3 2
- 8 4 0 1
---------
```

Take your pad and pencil and subtract the above problem from *left to right.* It will feel strange the first time, but your answer will come out right. If you feel at all uneasy about it, reassure yourself by doing it over in the way you are accustomed to working and note that the answer is the same.

Because working from left to right is a much harder adjustment to make in subtraction than it is in addition, do a few more examples in this way before going on to the complement techniques:

```
 1 6 8 3      3 3 6 5      5 9 7 0 1 2
 - 5 8 2    - 2 1 4 3    - 9 6 0 1 1
 -------    ---------    -----------
```

Just to make sure that you really have the idea, do them over again to see if your answers agree.

When we come to problems in which any digit of the smaller number is larger than the corresponding digit of the larger number, we face the situation handled in traditional methods by "borrowing." The relationship is really the reverse of the similar situation in adding two digits that go over ten, which traditionally calls for "carrying" but which we now handle by "recording." Just as we have substituted recording for carrying, we will now in subtraction throw out the concept of borrowing and substitute for it a new technique we call *canceling.*

Here is a situation in which you must borrow or cancel:

$$\begin{array}{r} 3\ 4 \\ -\ 2\ 7 \\ \hline \end{array}$$

Schoolbook thinking would approach this problem, from right to left, in this fashion: "7 from 14 (borrow the 1 from the 3) is 7. 2 from 3—no, we borrowed a 1 so it is now 2—2 from 2 is 0. Answer: 7."

Working from left to right in complement subtraction, our thinking is quite different. First, we glance at the first column and "see" 3 − 2 as 1. We put it down. There is a reason for this, so bear with the obvious wrongness of that 1 for a moment—you will see why. Then we glance at the second column and "see" 4 − 7 as *4 plus the complement of 7*—and *cancel* a ten.

The complement of 7 is 3. 4 plus 3 is 7. Put it down under the second column.

Keeping in mind that subtraction is just the reverse of addition, it should make sense that when subtracting you *add* a complement, just as when adding you *subtract* it. A full explanation comes later, but for the moment just remember that you are (in effect) doing addition in reverse and so your complements are added rather than subtracted.

Now we have used a complement, and when we use a complement in subtraction we must cancel a ten—just as when we use one in addition we must record a ten.

The method that makes possible our left-to-right working is that we cancel that ten in the *answer*—rather than "borrowing" it in the larger number. The technique for this is quite simple. We merely slash the 1 we put down under the first column:

$$\begin{array}{r} 3\ 4 \\ -\ 2\ 7 \\ \hline \not{1}\ 7 \end{array}$$

A slashed digit in the answer to a subtraction is a digit from which a ten has been canceled. In this particular case there is only one ten there—the ten of 17—so the answer is 7.

The general rule goes like this: To cancel a ten, slash the digit to the left in the answer. That digit is then reduced in value by one.

If there seems to be any confusion over the apparent interchangeability of the words "ten" and "one" here, reflect on the fact that each digit increases in importance by a factor of *ten* as it moves one place to the left.

Note the similarity of these answers to the last one, and follow the left-to-right process by which each was produced:

$$\begin{array}{r} 1\ 4 \\ -\ 7 \\ \hline \not{1}\ 7 \end{array} \qquad \begin{array}{r} 4\ 4 \\ -3\ 7 \\ \hline \not{1}\ 7 \end{array} \qquad \begin{array}{r} 7\ 4 \\ -6\ 7 \\ \hline \not{1}\ 7 \end{array} \qquad \begin{array}{r} 9\ 4 \\ -8\ 7 \\ \hline \not{1}\ 7 \end{array}$$

Now, however, keep in mind that a slashed digit is reduced in value by one—it is not wiped out entirely—and go through the development of these answers:

$$\begin{array}{r} 2\ 4 \\ -\ 7 \\ \hline \not{2}\ 7 \\ (1\ 7) \end{array} \qquad \begin{array}{r} 4\ 4 \\ -1\ 7 \\ \hline \not{3}\ 7 \\ (2\ 7) \end{array} \qquad \begin{array}{r} 7\ 4 \\ -3\ 7 \\ \hline \not{4}\ 7 \\ (3\ 7) \end{array} \qquad \begin{array}{r} 9\ 4 \\ -4\ 7 \\ \hline \not{5}\ 7 \\ (4\ 7) \end{array}$$

At this point the necessity for putting down that first digit at all, then slashing it and reading it as "one less" than it was before it was slashed, may be obscure. Its value and utility in working from left to right will become apparent when we get into longer problems with many columns, so make sure you understand the process thoroughly.

Why the Process Works

After visualizing the way complements function in adding, you have perhaps already seen the reason why the reverse should be true in subtracting. Let's go through a similar group of comparisons, however, to drive the point home.

Remember that group of ten-rung ladders. You are now

standing on the third rung of the fourth ladder. Your instructions are to step down exactly eight rungs. Where will you be standing then?

Obviously, you must drop down to the next ladder because you are only on the third rung of this one and you are to go down eight. If you descended a full ten rungs, you would then stand on the corresponding rung of that next-down ladder, or at the number 33. But you are to go down a number of rungs that fails by two (the complement of eight) to reach the corresponding rung—so you will be two rungs higher. You add the two, by which your eight-move fails to make ten, to the corresponding rung (three) and know that you will be on the fifth rung of the third ladder.

In simpler terms, 43 − 8 is 35. But you have arrived at this fact without ever subtracting 8 from (borrow) 3. Instead, you added the complement of 8 (2) to 3 to get the 5, and canceled a ten to reduce 4 to 3.

First, compare these two expressions:

$$\begin{array}{r} 4\ 3 \\ -\ 8 \\ \hline 3\ 5 \end{array} \qquad \text{or} \qquad \begin{array}{r} 4\ 3 \\ -\ 1\ 0 \\ \hline 3\ 3 \\ +\ 2 \\ \hline 3\ 5 \end{array}$$

Now see if you can feel the identity of these two expressions with the third, which describes our method of complement subtraction:

$$\begin{array}{r} 4\ 3 \\ -\ 8 \\ \hline 3\ 5 \end{array} \qquad \text{or} \qquad \begin{array}{r} 4\ 3 \\ -\ 1\ 0 \\ \hline 3\ 3 \\ +\ 2 \\ \hline 3\ 5 \end{array} \qquad \text{or} \qquad \begin{array}{rl} 4\ 3 & \\ +\ 2 & \\ \hline 4\ 5 & \text{and cancel a ten,} \\ & \text{which makes it} \\ & 35 \end{array}$$

Using complements instead of subtracting a larger digit from a smaller digit gives you not just one, but two major advantages in speed and accuracy. First, most of us find the process of adding easier than subtracting. Second, your thinking is restricted to the twenty easiest digit combinations and five complement pairs; you never deal at all in the pair $8 + 5$, for example, which is the digit-pair called for in our first expression $43 - 8$. Instead, your thinking is converted to the simpler pair $3 + 2$ by the use of a complement.

You also have a simple and highly automatic signal for the proper time to use a complement. In adding, it is when the two digits would add up to more than ten. In subtracting it is even easier. You use a complement whenever you would otherwise have to subtract a larger digit from a smaller.

Just remember, always, that subtraction is the reverse of addition. In adding, you subtract a complement. In subtracting, you add the complement—and always the complement of the digit being subtracted.

When adding, you record a ten every time you resort to a complement. When subtracting, you cancel a ten every time you use a complement.

Put the theory to use now by doing these four simple problems in the left-to-right method, using complements:

$$\begin{array}{r} 1\ 2 \\ -\ 4 \\ \hline \end{array} \qquad \begin{array}{r} 2\ 5 \\ -1\ 6 \\ \hline \end{array} \qquad \begin{array}{r} 3\ 7 \\ -2\ 8 \\ \hline \end{array} \qquad \begin{array}{r} 4\ 6 \\ -3\ 6 \\ \hline \end{array}$$

Easy as these are, they are designed to start you off with confidence in complement subtraction. Be sure to do them carefully and properly with the new technique.

The first example should develop like this: Nothing from 1 is 1. Put down 1. 4 is larger than 2, so do not subtract. Add complement of 4 (6) to 2. Put down 8, and immediately (before you forget) slash the 1 to cancel a ten. The answer is ~~1~~ 8, or 8.

Second: 1 from 2 is 1. Put down 1. 6 is larger than 5, so do not subtract. Add the complements of 6 (4) to 5 and put down 9. At once slash the 1 to cancel a ten. Answer, ~~1~~9. or 9.

Third: 2 from 3 is 1. Put down 1. 8 is larger than 7, so do not subtract. Add the complement of 8 (2) to 7. Put down 9. Immediately cancel a ten by slashing the 1. Answer, ~~1~~ 9, or 9.

The last example: 3 from 4 is 1. Put down 1. 6 is the same as 6. Nothing, or 0. No complement, no cancel. The answer is 10.

Perhaps the last one caught you. It was designed to. Complements only apply when we subtract a larger digit from a smaller. You will still subtract, about half the time, a smaller digit from a larger one or from one of the same value.

Why We Slash Digits

In the examples so far, it has really been a little childish to bother slashing digits in order to cancel tens. A fourth-grade schoolboy knows that 4 from 12 is 8. But you are exploring a new technique, a technique that applies not merely to 4 from 12 but also to 8,344,897 from 9,432,752. Learning to go through the proper steps is as important as learning to play the scales before tackling Chopin.

Play a few scales right now. First, make your complement-reaction just a little faster by "reading" the complements to these digits:

5 9 2 4 8 6 3 5 1 7

You will notice that in subtraction we now use *both* halves of each complement pair. We find it faster to use only the larger of each pair in addition, but you have to use all of them in subtraction. This is no problem, because there are still only five pairs. If you pause to wonder why we can pick which half of each complement pair we wish to use in adding, but have no choice in subtracting, notice that you can add $7 + 9$ or $9 + 7$ as you choose, but have no choice of complements in each of the two corresponding subtractions: $16 - 9$ or $16 - 7$.

Subtract the following examples from left to right. Put down on your pad or card every digit as you go along, even if it seems silly. This habit is important to your successful handling of longer and more complicated problems. Whenever you come to a larger digit from a smaller, add the complement of the digit to be subtracted to the digit you are subtracting from, and cancel a ten by slashing the digit to the left in the answer:

16	26	87	43	68
−8	−9	−78	−25	−42

One more point. A slashed 5 (5̸) is read as a 4, because the slash "borrows" or more properly "cancels" in the answer. But until this too becomes second nature, you may wish to rewrite answers before considering them finished. Remember that a slashed digit is reduced in value by one; then a subtraction answer that looks like this

6 8̸ 7̸ 2 4̸ 1̸ 9̸ 5

would be rewritten or would read like this

6 7 6 2 3 0 8 5.

After you have used this technique steadily for a few days, you will probably not bother to rewrite answers in this fashion. But until you have fully mastered the art of reading a slashed digit as one less than it was before the slash, you will profit by making sure you interpret such answers without error by rewriting them.

Take your pad now. Use it to cover the rewritten version of the following subtraction answer as you copy it in final form. Every slashed digit becomes the next digit smaller:

4 3̸ 8̸ 6 2̸ 0 9 6̸ 5̸ 2

After you have rewritten this answer, compare your version with the one that follows. If you got any of the digits wrong, it would be worth while to do it again.

Here is how your copied answer should read:

4 2 7 6 1 0 9 5 4 2

Now try these examples. Remember to work from left to right, use complements where indicated, and "borrow" by slashing the preceding digit in the answer:

4 8 3 9 0	1 4 9 2 5	7 3 6 1 8
− 3 9 9 3 9	− 3 7 5 7	− 7 2 8 0 9

By this time you should be finding it a little easier to work from left to right, and canceling tens in the answer—rather than "borrowing" in the larger number—should be beginning to feel natural. Once you become fully used to it you will find it far more natural and infinitely more foolproof than the older system.

It has been estimated that 80% of all mistakes in subtraction come from forgetting to borrow, or borrowing too much. Since we eliminate borrowing altogether, this method is by nature more accurate as well as faster.

Carrying Back Slashes

There is one more important element in this high-speed method of subtraction. This element is handling a slashed zero—∅.

A slashed zero is, like any other digit, reduced in value by one. Since it must have a digit to the left of it in the answer (or you could not subtract), then obviously the zero must become 9 – and the digit to the left of it must also be slashed, to reduce it in value by one (since you "borrowed" from it in order to get 09 from 10).

Consider this:

1̸ ∅

You would read the answer as 9. If the example were 2̸∅, you would read it as 19. In practice, particularly in a long problem, it is important to slash both digits. In reading that last 2̸∅, you would read 2̸ as 1, and ∅ as 9.

This may sound formidable, but it is really not as complicated as borrowing continuously to the left as you sometimes have to do in ordinary subtraction. Go through the steps in this example, and note where we start canceling tens:

```
 1 5 3 2 6 8
 − 5 3 2 6 9
 ───────────
 1̸ ∅ ∅ ∅ ∅ 9
  (9 9 9 9 9)
```

Follow each step carefully: Nothing from 1 is 1. Put down 1. 5 from 5 is 0. Put down 0. 3 from 3, and so on, gives you zeros until you come to the final column.

In the final column, 9 is larger than 8. Do not subtract.

Add the complement of 9 (1) to 8. Put down 9, and cancel a ten by slashing to the left.

The digit to the left is 0. Slash it. Whenever you slash a 0, you must go back and slash the digit to the left of it too. That next digit is also a 0, so you have to keep on slashing until you slash a digit that is not a zero.

This may still sound a little strange. If you have any lingering doubts, do the problem above in the old-fashioned, schoolbook fashion. You will find that you have to do precisely the same thing, but in the more complex, error-prone method of borrowing over and over for *each* subtraction.

Try two longer problems now. Remember, as always, to practice the new technique as you do them. Work from left to right. Subtract a smaller digit from a larger digit just as you do now. But do not subtract a larger digit from a smaller. Instead, add the complement of the larger digit to the smaller digit and slash left in the answer. If you slash a zero, remember to go back a step and slash the digit to the left of the zero too.

$$\begin{array}{r} 65437 \\ -43182 \\ \hline \end{array} \qquad \begin{array}{r} 11829635 \\ -9864713 \\ \hline \end{array}$$

The next chapter will carry you on to developing speed and accuracy at complement subtraction. Before you turn to it, however, let's cover another major advantage of adding and subtracting from left to right instead of from right to left.

Automatic Estimating

Any left-to-right method of doing arithmetic is self-estimating. Since you develop your answer from the left, the important end, you can always carry it exactly as far as you need for the accuracy you require and stop there.

Many of us have often tried to do this in the old-fashioned method of working when under pressure, but that is a backwards method and very difficult. Complement addition and subtraction does it automatically.

Suppose, for instance, you are production manager of a company making brass buttons. Your inventory as of the moment is 37,852 buttons. Today's orders total 16,965. The salesman selling to a large chain of stores calls to see how many buttons you could ship tonight on an emergency order. You must know, while he waits on the phone, about how many buttons you have.

Quick now: 1 from 3, 2. 6 from 7, 1. You have about 21,000 buttons. You have done merely the first two steps of your regular process in complement subtraction, instead of changing your method for estimating needs.

Suppose you need the next figure, too. 9 from 8. Add the complement of 9 (1) to 8: 9. Slash the digit to the left: 1̸. 20,900 buttons.

See how quickly and accurately you can give a three-digit estimate of the following subtractions:

```
  738,629      14,359,216
 -492,768     - 7,278,143
 --------     -----------
```

For estimating—as well as for many rounded-off computations—you simply ignore the relatively unimportant numbers to the right, and carry your subtraction just as far as you wish.

The automatic estimating feature applies just as much to complement addition as it does to subtraction. The only thing to beware of in adding is that whenever you stop, the next column could make a substantial change in your stopped-at digit. In subtracting, the next column can never affect your stopped-at digit by more than a reduction in value of one.

These two examples illustrate this point:

```
  19       90
  19      -39
  19      ---
  19
  19
  --
```

The illustrations are admittedly extreme. A first-column-only estimate of the addition would give a rough total of 50, while actually the real total is 95. A first-column-only approximation of the subtraction would be 60, while the real answer is 51.

The reason why the first digit of the addition can be changed in this case by 4, and the first digit of the subtraction is changed only by 1, is that you might be adding any quantity of numbers and any *two* of them can add up to more than ten—carrying back as much as ten for each two numbers. In subtraction, you never deal with more than two numbers and the maximum amount that can be canceled is one ten.

In subtraction, the safe approach is to work out your subtraction to one more digit than you really need, and round off. In adding, carry your addition *at least* one more place than you really need and assume that the final digit is raised by one for each two numbers you have added, then round off; or else carry it two digits beyond the accuracy required and round off.

Try one estimate in addition at this point. Give a rounded-off three-digit approximation of the following problem (The section immediately following takes up rounding off, in case you are not acquainted with the technique.):

```
248,396,211
783,210,650
151,899,000
976,482,722
382,000,000
334,836,255
-----------
```

If you worked this out to four digits and assumed the last digit would be raised by 3 (since you added six numbers), your working figures would be 2874 plus 3, or 2877. This you would round off to 1,880,000,000. If you went to five digits, they would be 28770. You would still round off to 2,880,000,000.

A properly rounded-off three-digit estimate can never at worst be more than one per cent wrong, incidentally, and more usually is restricted to no more than one-half of one per cent. The maximum error would be in an estimate of 100 when the accurate answer is 101. An estimate of 999, if properly rounded off, cannot be wrong by more than one-tenth of one per cent. Numbers in between have a maximum possible error that increases as the first digit decreases, from 9 to 1, but it cannot go over one per cent. This, once again, is because each digit becomes just one-tenth as important as it moves one place to the right.

How to Round Off

If anyone doesn't know how to round off, he has missed one of the greatest time- and energy-savers in modern business. Traditional accountants kicked and dragged their heels until they had worked with it a bit, then became its most enthusiastic supporters.

Rounding off simply means expressing any quantity to the nearest standard unit. The standard unit may be whatever you say it is. In the three-digit estimates you just did, we in effect determined that the standard unit would be one in which there could not be an error greater than one per cent.

The standard unit in a U. S. personal income-tax report is one dollar. $3.99 is rounded off to $4.00. $3.01 becomes $3.00. To become a little subtler, $3.51 becomes $4.00 and $3.49 becomes $3.00. The usual rule is to give away an even half, and call $3.50 an even $4.00.

Any other standard unit that makes sense for a particular situation can be adopted. The operating and financial statements of many companies are rounded off to even thousands. $357,800 is expressed on the statement as 358—with a note at the top of the report, of course, that all figures are in thousands of dollars. Smaller companies may round off to tens or hundreds of dollars. Very large corporations may even round off to hundreds of thousands or, for certain purposes, to the nearest million!

At the other extreme, there is an almost forgotten currency value in this country of one mil—a tenth of a cent. It was used primarily in state sales taxes, before sales taxes went up to much higher rates. Naturally, people working with quantities of mils soon learned to round off their reports—to the nearest cent!

The most accurate way to estimate in adding or subtracting, as we have said, is to work out your figures to one place more than the accuracy needed, and round off. If the extra (not needed) digit is 5 or more, raise the preceding digit by one before reporting the estimate. If the extra digit is 4 or less, leave the final significant digit alone.

The theory is that roundings-off tend to cancel each other out in practice. You will add half or less to just about as many numbers as those from which you subtract less than half. To the surprise of many old-line accountants and bookkeepers, several test-runs of complicated reports and statements proved this to be completely true. The error is hardly ever likely to be larger than a single rounding-off.

Review quickly now the three secrets of speed in subtraction, before going on to practice that speed. The three major secrets—in addition to the over-all speed-math secret of "seeing only the answer"—are

1. Subtract from left to right.
2. Never subtract a larger digit from a smaller. Instead, add the complement of the larger digit to the smaller digit and—
3. Cancel tens in the answer by slashing, rather than "borrowing" in the larger number.

5

BUILDING SPEED IN SUBTRACTION

CERTAIN parts of this book may seem repetitious. This is intentional. Repeating the basic points is the easiest and most painless form of review. Doing one essential exercise over several times—but not over and over in succession—is the most effective way to build the automatic response that is the foundation of high-speed mathematics.

Read the following line as if it were a sentence of words. But instead of words, read the complements of these digits:

4 7 2 8 5 1 9 6 5

Before going on to some necessary practice in complement subtraction, reinforce your understanding of the principle at work by describing in words completely different from any used in this book precisely what a complement is.

Now explain to yourself, as if you had never heard of the idea before, how you can subtract 7 from 12 by adding 3 plus 2—and doing something else in the answer. It might be a good idea to set up, on your pad, the three expressions $12 - 7$, $12 - 10 + 3$, and $12 + 3$ (cancel).

Your speed and ease with numbers will depend not only on how easily and automatically you "sense" these new techniques, but also on how easily and automatically you see only the answer to any digit combination. We will now go through the basic vocabulary of subtraction. It will not take very long, because the combinations are really the same ones you have already practiced for addition. They are all pairs you recognize at sight, but in this case one-half of each pair and the addition-answer are given, and you must respond with the missing number. $3 + 5$ is a pair you should be starting to read at sight as "8" instead of "$3 + 5$ is 8"; the same pair will show up here as $8 - 5$ (see 3) and $8 - 3$ (see 5).

Work for speed with these combinations. School yourself to think not about the digits you see, but only the answer. Where the bottom digit is smaller than the top digit, see only the answer. Where the bottom digit is larger than the top, work at seeing the result of adding the complement of the bottom digit to the top digit, and mentally slash an imaginary digit to the left in the answer. Therefore you "see" $6 - 7$ as "$3 + 6$ (9)—slash."

$$\begin{array}{cccccc}
7 & 4 & 1 & 6 & 8 & 4 \\
\underline{4} & \underline{7} & \underline{2} & \underline{4} & \underline{5} & \underline{3} \\
\\
3 & 6 & 2 & 9 & 1 & 2 \\
\underline{6} & \underline{2} & \underline{5} & \underline{3} & \underline{4} & \underline{7} \\
\\
5 & 9 & 1 & 5 & 6 & 4 \\
\underline{3} & \underline{5} & \underline{8} & \underline{8} & \underline{6} & \underline{1} \\
\\
6 & 3 & 8 & 9 & 5 & 7 \\
\underline{7} & \underline{2} & \underline{9} & \underline{7} & \underline{5} & \underline{2}
\end{array}$$

Just as in the practice tables in adding, every possible digit combination has been included in these sections. If you learn to read the answers to these without effort, you know you will never handle a single combination that you did not have a chance to practice.

See how quickly and automatically you can subtract these:

$$\begin{array}{cccccc}
8 & 7 & 4 & 3 & 6 & 4 \\
\underline{7} & \underline{3} & \underline{9} & \underline{3} & \underline{1} & \underline{6}
\end{array}$$

3	1	6	5	3	2
4	1	5	9	5	8
2	7	1	8	8	6
9	8	6	4	1	9
5	4	3	5	2	2
2	5	8	7	2	1

A certain amount of your speed at handling these must be pure habit, of course. There is no way to avoid developing the "automatic response" that only practice can bring. But the number sense at which you worked in Chapter II will be a substantial help here. The better you can visualize the relationships of numbers, the more quickly you will develop astonishing mastery of basic mathematical figuring.

Remember, too, that while this practice series shows 94 pairs (the 45 pairs, the 45 pairs upside down, plus the four hardest pairs repeated just to make it come out even), you need only be concerned with those twenty easiest combinations, plus the five complement pairs. Looked at this way, it should certainly be a reasonable task to master fully and automatically. What you are really doing, in effect, is learning to recognize those twenty easiest pairs whether they show up in simple smaller-from-larger form or disguised in complement applications.

Use your complements faithfully, and you will never deal with any combinations adding to more than ten—or subtract a larger digit from a smaller.

This finishes up all the possibilities:

1	9	3	5	1	3
9	2	9	6	7	1
4	9	8	9	6	3
2	8	6	9	8	7
6	5	4	8	1	4
3	4	4	3	3	8

7	2	7	6	9	5
6	3	5	2	1	1

7	2	9	8	9	2
1	4	4	8	6	6

1	7	7	9	7	4
5	7	8	8	9	9

That covers everything: 94 expressions of only twenty combinations, plus five complement pairs. Every problem you will ever face contains only these basic combinations, arranged in a different order. The only extra complication is your remembering to slash the answer-digit to the left whenever you use a complement. Even that is a far simpler system than trying to remember to "borrow."

Try this example. Be sure to use your pad:

$$\begin{array}{r} 4\,9\,3\,2\,7\,8\,5\,6\,1\,0\,2 \\ -\,3\,7\,8\,4\,6\,9\,6\,2\,0\,0\,8 \\ \hline \end{array}$$

A long subtraction, indeed. Yet you do it, step by step, in precisely the same way you would do your scales.

Use a clean page of your pad now and go through it once more. Compare the two answers. If they are not the same, you had better do it once again. Speed mathematics is useful only if it is also accurate.

Do these problems now, to help build your habits:

6842	13965	59	268
−5896	−8904	−49	−273

If you got an answer of any kind to that last one, take another look at it and bring your "number sense" to bear. There cannot be an answer, other than a minus one. It was put there to make sure you practice the reality, not an imitation.

Now do these:

4839	32115	12865	78
−4799	−20647	−234	−73

Each of these examples illustrates some variation of the pattern your left-to-right subtraction will form. Some of them require carrying back a slash to one or more preceding digits in the answer. Others may momentarily surprise you because they do not require the use of complements at all, and you will find no slashes whatsoever in your answer.

Do another group now:

9 3 6 3	1 3 2 5	4 6 5 2	3 8 6 4 3 3
− 8 8 7 4	− 3 2 3	− 3 8 7 9	− 4 7 3 8 7

The system works just as well, naturally, with dollars and cents. You can slash across a decimal point without hesitation, because as you move left each digit becomes ten times as important whether or not a decimal point appears between two digits. All the decimal point does is break the number into a whole quantity and a fraction. The digits retain precisely the same relative value right across the decimal point: ten times in value for each place a digit moves to the left.

In order to make sure that a decimal does not slow you up in your handling of canceled tens, work through these with your pad:

$ 2 . 9 8	$ 3 6 . 5 0	$ 1 7 5 . 0 0	$ 1 4 . 5 0
− 1 . 4 9	− 1 8 . 6 6	− 1 3 8 . 7 5	− 7 . 7 5

If you are rewriting your answers with each slashed digit reduced in value by one, then you have already had some good practice at reading such answers directly, without bothering to rewrite them.

Prove this to yourself by seeing if you can read the answer below as you would rewrite it, without pausing to figure out what each slashed digit should represent:

2 9̸ 6 8̸ 0̸ 4 9̸ 3

If you read through that like an expert, see if you can tell what is *wrong* with this answer:

3̸ 4̸ 6 0̸ 9 8̸

I would urge you not to skip over this. Unless one glaring error caught your eye in that answer, you would do well to review the last chapter on the subject of carried-back slashes. This is important—and will become even more important as you apply some of the elements learned so far to future sections of this book.

What Have You Learned

Unless you are unusually at home with numbers, or have a natural liking for them (which few of us do, although new mastery of any subject often brings enjoyment with it), now is a good time to pause and make sure everything covered so far is solidly entrenched.

You will profit most from this book if you take it in easy stages. Whenever a point seems a litle difficult to understand on one reading, go back and reread it once or twice. That same point is very probably one that will crop up again as something you will be expected to know thoroughly in new applications for multiplying and dividing. Take a pencil and your pad and doodle with the obscure point for a bit. See if you can set up different expressions of it, as we did for addition and subtraction involving complements. The idea is to visualize it as clearly as you can. In this way, you will understand the *why* as well as the *how*.

If you truly understand the why, I promise that you will never forget the how. Even if you did, you could easily reconstruct it—because you know why it works.

The next chapter will take up another major area of basic mathematics: multiplying. It is a fascinating and quite new approach, but do not tackle it until you feel completely comfortable with the complement, left-to-right methods of adding and subtracting. Between them, they account for 75% of the arithmetic used in the average business.

A final re-check would be in order now, to make sure your base is really solid.

First, find your own words to describe exactly what a complement is and how it works in adding. If you have trouble putting the theory into words, then set up the three expressions for adding 6 plus 9 on your pad in the same way we did before. Then do the same thing for subtracting 7 from 13.

Once you have lived a little longer with the idea of complements, they will seem to be the most natural and useful devices in the world. They are basic to the structure of our ten-based counting system. Yet, oddly enough, nobody had ever formalized their use for arithmetic until the Japanese found how much they simplified calculation on the abacus.

Skip ahead now to that last secret of extra speed in adding called grouping. Practice the technique briefly once again by grouping the following pairs at a glance as if they were together in a column you were adding and you wished to handle each pair as a single digit:

```
5   8   3   4   7   6   1   2
2   1   7   2   2   3   7   6
```

You are well on your way to mastery if your only reaction to that third group was "nothing, fold."

One more reading of the complements is now indicated. See the complements to the following digits as quickly and automatically as you can:

3 6 1 5 8 4 9 5 2 7

This brush-up on your basic vocabulary is not casual. It provides one more opportunity to drive the new habits a little more deeply into your mind, as well as to refresh your understanding of the principles at work.

Add the following example from left to right, using either the finger or line method of recording tens. Make sure to group whenever you can. Do it on your pad:

```
6 4 8 2
3 2 7 5
1 9 9 2
7 3 1 3
2 8 4 6
9 1 7 4
-------
```

It is entirely normal to hesitate a bit over some of the operations at this point. Do not worry if this happens to you. It takes quite a bit of living with any radically new methods before they become second nature. But if you have thoroughly understood each new idea and done each practice section conscientiously, you should have gone through each step without too much trouble.

Now re-check your left-to-right complement subtraction on the following problem. Use the slash method of canceling in the answer rather than borrowing in the larger number whenever you use a complement to "subtract" a larger digit from a smaller:

$$\begin{array}{r} 6,498,253 \\ -4,999,375 \\ \hline \end{array}$$

If you tackled each of these examples with dispatch and confidence, then you are ready for the brand-new method of no-carry multiplication.

6

NO-CARRY MULTIPLICATION

MULTIPLYING, according to the same estimates we mentioned before, averages about 20% of the figuring done in normal business.

But while it is used less than addition, multiplication is dreaded by more people, and done poorly or inaccurately by more people, than either addition or subtraction.

Perhaps this is because multiplication, particularly of one long number by another long number, becomes so fearfully complicated in comparison to the simpler process of adding or subtracting. Most of us have little trouble visualizing that adding ten lines of numbers may involve more work, but is really no more complex, than adding two lines of digits. But multiplying 2,958 by 165 is something that few of us can really "see" as a whole. We tackle it step by step, by pure rote, in inefficient and inherently slow traditional methods.

Our new way to multiply solves a large part of this. It involves three secrets. Two of them you already know from adding and subtracting. The third is brand-new.

The first secret is the one we use throughout this book: work from left to right. Tackle 165 as 165, not as 5, 6, 1. The less we have to use methods that violate the plain common sense of the way we normally read numbers, the better off we are. Our number sense becomes sharper instead of becoming dulled by backward absurdities.

Working from left to right also makes our new method of multiplying a self-estimating system, just as our left-to-right addition or subtraction is.

There is no simple way to work from left to right in the classical method of multiplying. But our no-carry method works just as easily from left to right as from right to left, so you will find it natural to work in the proper direction.

The second secret, again, is the same as its equivalent secret in adding or subtracting: "see" only the answer, combining digits at a glance. This is simply a matter of practice, but chances are you have already had more practice at this than you had for adding. Most schools spend far more time on drill in multiplication tables than they do on drill in addition and subtraction tables. So you are probably closer to mastering this step in multiplication than you were for the two earlier processes.

If you were taught in the standard way, however, you would do well to begin practicing the deletion of the slow-down steps taught in school. Instead of reading the example

$$\begin{array}{r} 4 \\ \times\ \underline{7} \end{array}$$

as "4 times 7 is 28," make a conscious effort to look at it and think only "28." You do not look at "me" and think "*m* and *e* is 'me.' "

In fact, the entirely new way to multiply involved in the third step will bring up quite a different way of looking at 4×7. You will never, oddly enough, think the whole product at all, but only half of it at a time.

The third secret is the new method. It is radically different from the traditional way to multiply because you never have to "carry." The greatest trouble with standard multiplication, and the greatest source of errors, is carrying. It is very much like the difficulty in "borrowing" in subtraction. Either you forget to carry, or carry twice, or carry the wrong figure—and wind up hating numbers.

The no-carry method of multiplying works without remembering to carry at all. It may look a little strange at first, but once you try it a few times you will get the idea.

The easiest way to approach this method is to take apart a sample multiplication and see what makes it tick. Make sure you fully understand every step of this, because once you understand why the system works as it does you will find it very easy to use. If you simply try to learn the technique by rote, however, it will always seem complicated.

Let's take this multiplication apart:

$$\begin{array}{r} 4\ 7 \\ \times\ 8 \\ \hline 5\ 6 \\ 3\ 2\ \ \ \\ \hline 3\ 7\ 6 \end{array}$$

Look at the answer, digit by digit, and see how it really develops.

The first digit, 3, is simply the *left-hand* (tens) digit of 4 times 8—32.

Look back at the example and note this. The first digit of this answer is merely the tens, or left-hand, digit produced by multiplying the first digit of the number multiplied by the multiplier.

The second digit is a little more complicated. This 7 is the sum of two other digits. It is the sum of the *right-hand* (units) digit of the multiplication we just examined—4 times 8—and the *left-hand* (tens) digit of 7 times 8. The right-hand digit of 4 times 8 is 32, or 2. The left-hand digit of 7 times 8 is 56, or 5. 2 plus 5 is 7—the middle digit in our answer.

Look back at the example again to make sure this is completely clear. Read the above explanation again if you need to.

If you remember our earlier comments about left-to-right working, in which we pointed out that each digit increases in value by a factor of ten as it moves one place to the left, then you can see why the middle digit in this answer is the sum of the *unit* part of the 4 times 8, and the *tens* part of the 7 times 8. It is because the 4 in 47 is really ten times 4 because of its position—or 40.

The last digit in this answer is 6. This is simply the *right-hand* (units) digit of 7 times 8—56.

This is a new way of looking at multiplication for most people. Get it clear now, and everything that follows will fall into place naturally and easily.

Now let us try multiplying those same numbers left to right in the new method, using the understanding above of how the answer really develops. If the method seems unclear at any point, re-check the explanation above.

$$\begin{array}{r} 4\ 7 \\ \times\ 8 \\ \hline \end{array}$$

Step one: Look at 4 × 8 to see only what the left-hand (tens) digit of the product will be. In other words, is 4 × 8 in the teens, twenties, thirties, forties, or what?

4 × 8 is in the 30's. The tens digit of this pair is 3. For the moment, you do not care what the right-hand, or units, digit is. All you care about is the 3.

For the first digit of your answer, put down that 3:

3

Step two: Now look at 4 × 8 to see what the right-hand, or units, digit of this pair is . . . what the full product of 4 × 8 "ends in." The units digit is 2, the 2 of 32. Remember that 2 for just an instant while you look at 7 × 8 to see what its left-hand, or tens, digit will be. 7 × 8 is in the 50's. Add this 5 to the remembered 2 and put down the total as the second digit of your answer. 2 plus 5 is 7, so your answer now looks like this:

3 7

Step three: Look at 7 × 8 again to see only what digit the product ends in. The right-hand, or units, digit of 7 × 8 is 6—the 6 of 56. Put it down as the last digit in your answer:

3 7 6

Pause here for a moment to let this sink in. It is just as shocking an idea in its own way as is the idea of complements for adding and subtracting, and just as useful. But, as with complements, you need a bit of time to adjust to the thought.

There is one point in step two when you must remember one digit while "seeing" another one to add to it. Check the traditional process taught in school, however, and you will find that you had to juggle three digits at this point. You had to carry the 5 from 56 while noting the 32, then remember the 3 from 32 while adding the 2 and 5 and putting down 7. After that, you had to remember to put down the 3 from 32. The new no-carry method is at least one-third simpler—and produces the answer from left to right as well.

Here is another run-through to reinforce your grasp of this method:

$$\begin{array}{r} 8\ 3 \\ \times\ 9 \\ \hline \end{array}$$

Step one: 8 × 9 is in the 70's. Write down 7 as the first digit of your answer:

7

Step two: 8 × 9 ends in 2. Remember 2. 3 × 9 is in the 20's. Add the remembered 2 and the 2 from the 20's and put down 4:

7 4

Step three: 3 × 9 ends in 7. Put down 7 as the last digit in your answer:

7 4 7

If any element along the way does not seem to make sense, go through the three steps again with pencil and pad. This is really an incredibly simple idea, but it is vastly different from the way we were taught to work with numbers.

Now we will try one more, adding another digit. This means simply that we shall do step two twice. More properly, steps "one" and "three" are special steps for the extreme left and right digits of the number multiplied. Step "two" is the step done for every pair of digits across the number multiplied; once for a two-digit number, twice for three digits, and so on.

Here is how it works with a three-digit number:

$$\begin{array}{r} 5\ 3\ 2 \\ \times\ 7 \\ \hline \end{array}$$

One: 5×7 is in the 30's. Put down 3:

3

Two (1): 5×7 ends in 5. Remember 5. 3×7 is in the 20's. Add 5 and 2. Put down 7:

3 7

Two (2): 3×7 ends in 1. Remember 1. 2×7 is in the 10's (teens). Add 1 and 1. Put down 2:

3 7 2

Three: 2×7 ends in 4. Put down 4:

3 7 2 4

That is the basic system. It is that simple, and that revolutionary. If there had been twenty digits in the number multiplied, you would simply have repeated step two until you got to the end.

Get out your pad, open to a clean page, and go through the steps exactly as described for the following example. Do not try it on other random problems yet, however, because there are two special techniques for special cases yet to be revealed.

$$\begin{array}{r} 4\ 7 \\ \times\ 4 \\ \hline \end{array}$$

After you have done this, check your answer by the usual method of multiplying. If it checks out, good. If not, go back through the steps and see where you went wrong.

If it still does not come out right, compare your working with this description of the proper steps:

Step one: 4×4 is in the 10's. Put down 1.

Step two: 4×4 ends in 6. 7×4 is in the 20's. Add 6 and 2. Put down 8.

Step three: 7×4 ends in 8. Put down 8. The final answer is 188.

Now for the two special cases. Both are important, because examples involving them will crop up repeatedly in your work with numbers.

How to Handle Zeros

Sometimes, in going through the no-carry multiplying system, you will match a pair of digits whose product is less than ten. It might be 3 × 2. This product is 6. There is no left-hand, or tens, digit at all. In effect this product is in the "zeros."

For this system, however, you must use a left-hand digit. Otherwise the answer will not come out right. So no-carry multiplication always depends on using a left-hand digit even if that digit is zero.

When you come across 3 × 2, you will consider it in the zeros, just as 3 × 4 is in the 10's, and 3 × 7 is in the 20's.

The reason for keeping this in mind is that your left-hand and right-hand product digits are essential to keeping your imaginary "carries" in proper order. Later, when we come to working with two or more digits in the multiplier, you will find them important for keeping your columns in line too. This is really no more difficult than remembering to put down the zero in 5 × 6 when working from right to left, and performs basically the same function.

Suppose, for instance, you faced this example:

$$\begin{array}{r} 5\ 1\ 4 \\ \times\ 7 \\ \hline \end{array}$$

Step one: 5 × 7 is in the 30's:

3

Step two (1): 5 × 7 ends in 5. 1 × 7 is in the *zeros.* 5 plus zero is 5:

3 5

Step two (2): 1 × 7 ends in 7. 4 × 7 is in the 20's. 7 plus 2 is 9:

3 5 9

Step three: 4 × 7 ends in 8:

3 5 9 8

One other important point about products whose left-hand, or tens, digits are in the zeros should be kept in mind. Get in the habit of putting down a zero as the first digit of the answer if this is what the problem produces. It is not essential for one-line answers such as those in the above examples, but it is absolutely essential to getting two-line answers lined up properly.

This is what I mean:

$$\begin{array}{r} 1\ 6 \\ \times\ 4 \\ \hline \end{array}$$

Step one: 1×4 is in the zeros. Put down 0:

Step two: 1×4 ends in 4. 6×4 is in the 20's. Add 4 and 2. Put down 6:

0 6

Step three: 6×4 ends in 4. Put down 4:

0 6 4

Your answer is merely 64. The zero in front of it does not change its value. But when you come to multiplying by numbers of two or more digits, you will see the necessity of this technique. It is for precisely the same reason, as we said a page back, that in the traditional method you put down the zero of thirty or forty at the right of the answer in a two-line multiplication.

But since this is a new way of doing things, be sure to get into the habit of doing it this way whenever the problem works out like this. Try it on these two samples. Use your pad:

$$\begin{array}{r} 4\ 9 \\ \times\ 2 \\ \hline \end{array} \qquad \begin{array}{r} 3\ 8\ 6 \\ \times\ 2 \\ \hline \end{array}$$

Be sure to do these. Simply reading through practice examples, intending to do them later, will not teach you how to do speed mathematics. Theory and practice go hand in hand.

Check your results and the steps you went through in the two samples above against this explanation:

First sample. Step one: 4×2 is in the zeros. Put down 0.

Step two: 4 × 2 ends in 8. 9 × 2 is in the 10's. 8 plus 1 is 9. Step three: 9 × 2 ends in 8. Answer: 0 9 8.

Second sample. Step one: 3 × 2 is in the zeros. Put down 0. Step two (1): 3 × 2 ends in 6. 8 × 2 is in the 10's. 6 plus 1 is 7. Step two (2): 8 × 2 ends in 6. 6 × 2 is in the 10's. 6 plus 1 is 7. Step three: 6 × 2 ends in 2. Answer: 0 7 7 2.

If you neglected to put down the zeros in front of these answers, do (for the sake of your swift mastery of two-digit multipliers) go back and do them properly now. Simple repetition, pencil in hand, means a great deal in getting accustomed to new techniques such as these.

The new way of looking at half-products may come a little hard at first. You were taught to think "6 times 8 is 48." Now, in two separate steps, you are learning to look at 6 times 8 and (for one step) think only "40's," then (for another step) think only "8." Don't worry about that part yet. It is really quite a simplification of the multiplication tables, and there is some practice ahead to help give you the knack.

Before going on to the final step in no-carry multiplying, get a firmer grip on the steps so far by doing these two examples. Turn to a clean page of your pad and try your teeth on these:

```
9 3 6        7 4 9
  × 4          × 6
-----        -----
```

We will not go through the explanation of these in detail. Do them thoughtfully and carefully, working at this point for full understanding and accuracy rather than speed. Speed will follow because you are now working in what is essentially a simpler and more logical manner.

Once you have done the two examples, check your results by repeating the two problems according to your old method. If the answers check out, fine. If not, study the steps in detail to find out where you went wrong.

Now for the final step.

Recording Tens

So far, all our examples have been carefully selected to avoid one special situation that is really more complicated in the traditional method than it is in no-carry multiplying. But the situation does need a technique to handle it, and we have a simple and automatic one.

This example will demonstrate the special situation. Go through it step by step and find the new problem:

$$\begin{array}{r} 8\ 9 \\ \times\ 6 \\ \hline \end{array}$$

Step one: 8 × 6 is in the 40's:

$$4$$

Step two: 8 × 6 ends in 8. 9 × 6 is in the 50's. 8 plus 5 is—

STOP! You cannot put down a single digit standing for the sum of 8 and 5. This goes over ten. In our new way to add, we do not even try to add them. Instead, we subtract the complement of 8 (2) from 5 and put down 3:

$$4\ 3$$

But this is not quite right. When we use a complement, we must also record a ten. How do we record a ten here?

One of the secrets of this simplified mathematics is that we let the tens take care of themselves. We record them in adding, or cancel them in subtracting. We never, never try to remember them. That would be inefficient.

In multiplying, then, we will simply use the same written symbol we use in adding. We will underline. In this case, an underline will raise the value of the underlined digit by one—just as, in subtracting, a slash reduces the value of the slashed digit by one. Since the underline is in effect *carried back* from the 3 (the underline represents the 1 in 13, which is one place to the left), we will underline the digit one place to the left.

So our answer now looks like this:

$$\underline{4}\ 3$$

Step three: 9 × 6 ends in 4:

$$\underline{4}\ 3\ 4$$

Since the underline raises the value of the underlined digit by one, we read our answer like this:

5 3 4

Is this correct? Check the problem and see. Just as important, or even more important, see if the logic of it is clear.

If this seems the least bit complicated, review in your mind the schoolbook approach to this problem. Here is the thinking you were instructed to do: "6 × 9 is 54. Put down the 4, carry the 5. 6 × 8 is 48. We carried a 5. 8 plus 5 is 13. Put down the 3. Carry one from the 13. Add the 1 of the 13 to the 4 of 48. 1 plus 4 is 5. Put down 5." Which, once you are equally familiar with both approaches, is really simpler?

Let's go through this new process once more in detail:

$$\begin{array}{r} 4\ 6\ 8 \\ \times\ 3 \\ \hline \end{array}$$

Step one: 4 × 3 is in the 10's:

1

Step two (1): 4 × 3 ends in 2. 6 × 3 is in the 10's. 2 plus 1 is 3:

1 3

Step two (2): 6 × 3 ends in 8. 8 × 3 is in the 20's. 8 and 2 are complements. Zero, record:

$1\ \underline{3}\ 0$

Step three: 8 × 3 ends in 4:

$1\ \underline{3}\ 0\ 4$

Until you are thoroughly accustomed to reading slashed and underlined digits accurately without hesitation, it is good practice to rewrite such answers:

$1\ \underline{3}\ 0\ 4$ becomes

1 4 0 4

Now try doing a problem that involves this situation on

your own. Write your answer, left to right, before going on. Then check your steps against the explanation that follows:

$$\begin{array}{r} 7\ 8 \\ \times\ 8 \\ \hline \end{array}$$

Here is the way the no-carry method works on this problem:

Step one: 7 × 8 is in the 50's:

$$5$$

Step two: 7 × 8 ends in 6. 8 × 8 is in the 60's. Complement of 6 (4) from 6 is 2, and record the ten:

$$\underline{5}\ 2$$

Step three: 8 × 8 ends in 4:

$$\underline{5}\ 2\ 4$$

Rewrite this answer as 624, and you are done.

Sometimes, too, your recorded ten will affect a first-place zero. Here is such a case:

$$\begin{array}{r} 1\ 9 \\ \times\ 8 \\ \hline \end{array}$$

Work this one out yourself, being sure to put down a zero if the left-hand, or tens, digit of any of the products is in the zeros, and see if this changes as you go through the full answer.

Work it out now on your pad.

If you did each step correctly, your answer should look like $\underline{0}52$, which you rewrite as 152.

Do the next two problems on your pad. Be sure to go through exactly the steps we have been demonstrating, and check your answers to make sure they are right. If you hesitate too much, or come up with a wrong answer on two or three tries, then a review of the last few pages is in order.

$$\begin{array}{r} 9\ 3\ 6 \\ \times\ 7 \\ \hline \end{array} \qquad \begin{array}{r} 4\ 8\ 5 \\ \times\ 6 \\ \hline \end{array}$$

How Complements Help

By this time, you have surely noticed a surprising and delightful fact: complement addition is a tremendous aid to no-carry multiplication. You need have no worry about remembering when to record a ten. The occasion is signaled to you automatically. You record a ten every time you use a complement or add to ten, and that is all. One goes with the other.

Whenever you use a complement, of course, this also gives you the digit to enter in the answer more quickly and accurately than would trying to add (say) 9 plus 9 and getting 18—of which you would have to put down the 8 and carry the ten. Working the new way, you think simply "8, record."

It is almost foolproof, once it becomes a habit. No-carry multiplication is the closest possible approach to the secret of the soroban: to make as much of the operation as possible mechanical, so you can give your attention to the digit-by-digit sums and products and divisions without worrying about carrying and holding numbers in your mind. Since you are released from much of the mental labor of ordinary mathematics, you can concentrate on the single most important skill needed to handle this quickly and easily: your ability to "see" 6 × 7 as "40's" and "ends in 2" without hesitation or effort. The next chapter will go more fully into this.

Two-Digit Multipliers

No part of no-carry multiplication changes when we approach multipliers of two or more digits. It would be theoretically possible to produce the answer to two- or three-digit multipliers in one operation, but this means juggling four digits in your mind at once. This, for most of us, is impractical. Some "short-cut" mathematics books do urge this method, but it is really going in precisely the wrong direction for true speed. Unless years of practice go into them, the methods for producing the answer to 59 times 38 in one operation, on one line, are more apt to get mixed up and give a wrong answer than to speed up your results.

You now know how to produce a left-to-right answer to any multiplication by one digit with greater speed and accuracy, as well as ease. Let us stick to this head start, and do longer problems in the simplest and fastest way. We will use a different line for each digit in the multiplier. We will arrange these lines, however, in the reverse of the classical system. Start your top line with the *left* digit of the multiplier, and put each following line one place to the right.

This method is both easier and faster once you get the hang of it. It keeps you working from left to right—which is more natural, and also makes the system self-estimating.

Watch this demonstration:

$$\begin{array}{rr} & 2\ 6 \\ \times & 9\ 8 \\ \hline \end{array}$$

Step one: 2 × 9 is in the 10's:

$$1$$

Step two: 2 × 9 ends in 8. 6 × 9 is in the 50's. Complement of 8 (2) from 5, and record:

$$\underline{1}\ 3$$

Step three: 6 × 9 ends in 4:

$$\underline{1}\ 3\ 4$$

Now, for the second line. This we get by multiplying the 8 by each digit in turn of the number multiplied, and we place the answer one place to the *right* (we work always from left to *right*):

Step one: 2 × 8 is in the 10's:

$$\begin{array}{ccc} \underline{1} & 3 & 4 \\ & 1 & \end{array}$$

Step two: 2 × 8 ends in 6. 6 × 8 is in the 40's. 6 and 4 are complements. Zero, record:

$$\begin{array}{ccc} \underline{1} & 3 & 4 \\ & \underline{1} & 0 \end{array}$$

Step three: 6 × 8 ends in 8:

$$\begin{array}{cccc} \underline{1} & 3 & 4 & \\ & \underline{1} & 0 & 8 \end{array}$$

Now you simply add these lines from left to right. The easiest way to handle the recorded tens is to add each underline as one, rather than rewrite the two lines before adding:

$$\begin{array}{cccc} \underline{1} & 3 & 4 & \\ & \underline{1} & 0 & 8 \\ \hline 2 & 5 & 4 & 8 \end{array}$$

This should be clear through every step. If not, recheck your understanding of the preceding pages.

There is one special point to watch carefully. You recall the stress we put on putting down a first digit for each line, even if that first digit happens to be a zero. If you forget to do this, your lines for each digit in the multiplier will get out of order and your answer will be wildly wrong. Notice how it works in this example:

$$\begin{array}{cccc} & 2 & 4 & \\ \times & 3 & 6 & \\ \hline 0 & 7 & 2 & \\ & 1 & 4 & 4 \\ \hline & 8 & 6 & 4 \end{array}$$

If you had not put down the zero in front of the 7 in that first line, you might very possibly have lined up the two partial-products like this:

$$\begin{array}{cccc} 7 & 2 & & \\ & 1 & 4 & 4 \\ \hline 7 & 3 & 4 & 4 \end{array}$$

Not only is the answer absurdly wrong, but it is the type of error that is infernally hard to catch. You might do the problem over several times, get every line right, and still get the wrong answer. So watch, very carefully, your first-place zeros.

Go through the process once yourself. Doing is the secret of remembering:

$$\begin{array}{r} 8\ 6 \\ \times\ \underline{1\ 3} \end{array}$$

Cover the solution below with your pad while you work out this example.

If you remembered that you always put down a left-hand digit even if it is a zero, then your answer looked like this:

$$\begin{array}{r} 8\ 6\ \\ \times\ \underline{1\ 3}\ \\ 0\ 8\ 6\ \\ \underline{2\ 5\ 8} \\ \underline{0}\ \underline{0}\ 1\ 8 \end{array} \text{ or } 1{,}118$$

Note how we handled the two recorded tens in adding the lines of partial answers. You are already learning to read underlined digits as increased in value by one. Occasionally, in very complex multiplications with three, four, or more digits in the multiplier you will need to record more than one ten for the same place. That is, sometimes you will need to add two or three to a digit already put down as you work from left to right.

In any multiplication likely to become this involved, you may wish to handle the addition of partial answers just as if it were a major addition problem, and record tens on your fingers to be noted beneath each digit to the left. But multiplication seldom becomes this complex. In most cases, you will find it quicker and easier to use two or even three underlines when you need to.

Now it is time to handle a couple of two-digit multiplications entirely on your own. Cover up the answers below with your pad while you use it to work out the answers. Use your new techniques exclusively.

$$\begin{array}{r} 7\ 8 \\ \times\ \underline{8\ 2} \end{array} \qquad \begin{array}{r} 4\ 9 \\ \times\ \underline{4\ 3} \end{array}$$

After you have finished, compare your working with these answers:

$$\begin{array}{r} \underline{5}\ 2\ 4\ \\ \underline{1\ 5\ 6} \\ 6\ 3\ 9\ 6 \end{array} \qquad \begin{array}{r} 1\ 9\ 6\ \\ \underline{1\ 4\ 7} \\ \underline{1}\ \underline{0}\ 0\ 7 \end{array}, \text{ or } 2{,}107$$

Larger Multipliers

When you get into three- and four-digit multipliers, you simply follow the steps you already know. They are no different in kind, only in degree. Just add one more line for each new digit in the multiplier, stepping one place to the right for each line, and remember to put down a left-hand digit even if it is a zero in order to keep the lines in proper order.

There is no magic to getting the right answer for a long multiplication problem. There are two ways of getting very rapid, quite accurate estimates—the slide rule, and the self-estimating feature of no-carry multiplication—but for a full, complete answer you simply have to go patiently through all the steps. Those steps are made more natural and easier as well as quicker by this method.

Watch the step-by-step development of the answer to the following example. You should be able to understand why each new digit appears without trouble. If anything does not seem clear, review the last few pages and try again. It might be a good idea to follow with pencil and pad, too.

$$\begin{array}{r} 976 \\ \times\ 438 \\ \hline \end{array}$$

1. $\begin{array}{l} 3 \end{array}$

2. $\begin{array}{ll} 3 & 8 \end{array}$

3. $\begin{array}{lll} 3 & \underline{8} & 0 \end{array}$

4. $\begin{array}{llll} 3 & \underline{8} & 0 & 4 \end{array}$

5. $\begin{array}{llll} 3 & \underline{8} & 0 & 4 \\ & 2 & & \end{array}$

6. $\begin{array}{llll} 3 & \underline{8} & 0 & 4 \\ & 2 & 9 & \end{array}$

7. $\begin{array}{llll} 3 & \underline{8} & 0 & 4 \\ & 2 & 9 & 2 \end{array}$

8. $\begin{array}{lllll} 3 & \underline{8} & 0 & 4 & \\ & 2 & 9 & 2 & 8 \end{array}$

9. $\begin{array}{lllll} 3 & \underline{8} & 0 & 4 & \\ & 2 & 9 & 2 & 8 \\ & & 7 & & \end{array}$

10. $\begin{array}{lllll} 3 & \underline{8} & 0 & 4 & \\ & 2 & 9 & 2 & 8 \\ & & 7 & 7 & \end{array}$

11. $\begin{array}{lllll} 3 & \underline{8} & 0 & 4 & \\ & 2 & 9 & 2 & 8 \\ & & 7 & \underline{7} & 0 \end{array}$

12. $\begin{array}{llllll} 3 & \underline{8} & 0 & 4 & & \\ & 2 & 9 & 2 & 8 & \\ & & 7 & \underline{7} & 0 & 8 \end{array}$

Now add, from left to right of course:

$$\begin{array}{cccccc} 3 & \underline{8} & 0 & 4 & & \\ & 2 & 9 & 2 & 8 & \\ & & 7 & \underline{7} & 0 & 8 \\ \hline \underline{3} & \underline{1} & \underline{6} & 4 & 8 & 8 \end{array}$$

or

$$4\,2\,7,4\,8\,8$$

Go through the mental processes in your mind, making sure that you too would put down each new digit as it appears in the step-by-step unfolding of this answer. Note especially how the underlines are handled.

Now try a couple of three-digit multiplier problems on your own. Work from left to right, in the no-carry method, and remember to put down a left-hand digit for your first product in each line even if it is a zero:

$$\begin{array}{r} 4\,6\,7 \\ \times\,1\,3\,9 \\ \hline \end{array} \qquad\qquad \begin{array}{r} 8\,7\,3 \\ \times\,5\,4\,6 \\ \hline \end{array}$$

You will find the detailed working of these two examples at the end of this chapter. Get your pad now, though, and go through them to the end before reading on. Save your working for a check against the solutions to come later.

Automatic Estimating

One of the beautiful features of left-to-right, no-carry multiplication is the way it produces quick estimates. It is as fully automatic in this respect as is left-to-right addition or subtraction.

There is no easy and accurate way of doing this with traditional multiplication. Yet it is built right in, at no extra cost, to any left-to-right system.

You can get a two-digit estimate in a twinkling. You can get a three-digit estimate (which equals the accuracy of almost any slide rule) while the man with the slide rule is still getting out his "slip stick" and setting it.

This is not a criticism of the slide rule. If you must do a

great deal of multiplying and dividing and are satisfied with rounded-off answers—which the slide rule provides by its very nature—then it is well worthwhile getting one and learning how to use it. It is not hard. But do not pass up this estimating short cut even if you have a slide rule, because the system is both useful and impressive. It also works when your slide rule is somewhere else.

The technique for estimating with no-carry multiplication to any required degree of accuracy is simply this: Multiply as far as you have to and stop. Raise the last digit by one for each two digits in the multiplier.

Suppose you face a really formidable multiplication such as the cost of 53,926 items at $48.75 each. You must give a rapid approximation to three digits.

All you need to do is quickly scrawl each part of your new no-carry multiplication as far as three digits from the left. Here is how you do it:

```
  5 3 , 9 2 6
× 4 8 . 7 5
  -----------
  2 1 5
    4 2
      3
  -----
  2 5 0
    -
```

So far, you have 260. There are four digits in the multiplier—count 4, 8, 7, 5—so raise the 0 by 2. Now you have 262.

A slide rule would not do any better. Carry the multiplication further, if you wish, and see how close we are.

Does this mean $262,000 or $2,620,000? One simple rule gives you an unfailing answer to this question. Your answer has exactly as many digits as the total of the two numbers multiplied. Just add the digits in these two numbers, and figure on that many in the answer. Special note: If the first digit of the answer is a zero (the first digit of the first line of partial answers), this must be counted too.

In the above estimate, your answer is $2,620,000. Try working it out and see—keeping in mind that two of the digits in the multiplier are behind the decimal point and therefore are a fraction.

An estimate of

$$
\begin{array}{r}
2,\ 4\ 6\ 8 \\
\times\ 2\ 8 \\
\hline
0\ 4\ 9\ 2 \\
1\ 9\ 6 \\
\hline
0\ \underline{5}\ 8\ 8
\end{array}
$$

would have five digits in the final answer, or 68,900. The total number of digits in number multiplied and multiplier is six, but in the answer one of that total is lost in the initial zero.

Two other special points are interesting in this matter of estimating. First, note that you work your answer out to only as many places as you need and, in order to do it, you work out each partial answer to this same number of places *starting at the top left*. For a three-digit estimate, you will have three digits in the top partial answer, two in the second, and one in the third. If the first line begins with a zero (as in the one above) then you will go to four digits in the first line. Should your multiplier have twenty digits in it, you would ignore all but the first few.

Perhaps you wonder why you raise the last digit by one for each two digits in the multiplier. Check back to the section on estimating in the chapter on subtraction, and you will find a very similar rule. The reason is this: The average of any random number of digits including 0 is 4½. The average for each two lines in addition is therefore 9—*plus* the likelihood of tens recorded (or carried back to this column) from the column to the right, at the rate of about one for each 2½ lines. The best average for estimating, then, is to increase your final digit by one for each two lines in the addition. And the number of lines in the final addition of a multiplication problem is determined by the number of digits in the multiplier: one line of partial answer for each digit.

So raise the final digit of your estimate by one for each two digits in the multiplier. Forget any extra digits, and count five as two, seven as three.

Practice estimating these two problems accurate to three digits. Use pencil and paper. Remember to raise the last digit of your estimate in the way described above, and to count the digits in both numbers and use this total as the number of digits in your answer—including an initial zero if it appears in the top line of partial answers.

Do these now:

```
  4 6 , 8 3 2        8 3 6
  × 8 , 9 6 3      × 5 1 7
  -----------      -------
```

The estimates of these appear at the end of the chapter. Do them yourself, though, before you look.

• • •

Here are the answers to the two three-digit problems you were asked to work out on page 78. Compare them with your solutions:

```
   4 6 7               8 7 3
 × 1 3 9             × 5 4 6
 -------             -------
   0 4 6 7             4 3 6 5
     1 3 0 1             3 4 9 2
       -                   
       3 1 0 3             4 2 3 8
       - -                 -
 -----------         -------------
     5 4 9 1 3         4 6 5 6 5 8
     -                   - -
        or                 or
     64,913              476,658
```

And here is the way we estimate to three-digit accuracy the two examples at the top of this page:

```
 4 6 , 8 3 2             8 3 6
 × 8 , 9 6 3           × 5 1 7
 -----------           -------
 3 6 4                   4 1 8
   -                       0 8
   3 1                       5
   -                   -------
     2                   4 1 1
 -------                   =
 3 1 7
 -
```

Add 2 (4 digits in multiplier): 4 1 9

Nine digits:

419,000,000

Note especially the two underlines, meaning two recorded tens.

Add 1 (3 digits in multiplier): 4 3 2

Six digits:

432,000

The next chapter will help you to develop greater familiarity and speed with these techniques. If you feel that everything in this chapter is completely clear, go on ahead. If not—review.

7

BUILDING SPEED IN MULTIPLICATION

YOU recall that we stated three basic secrets for speed in multiplication:

First, work from left to right (possible only with this system).

Second, "see" the result of each multiplication of two digits, rather than the problem.

Third, use the no-carry method.

The second point is the one that obviously requires the most practice. The foundation of all your speed is the easy, natural, painless use of the no-carry system—but the way to make it easy and painless is to make as automatic and unthinking as possible the process of "seeing" 8 × 7 as "50's" and "ends in 6."

Your job now is to go over these half-products enough times to make the automatic response a habit. Since you undoubtedly got far more drill in the multiplication tables than you did in addition and subtraction tables, learning to see each product as only the left-hand or right-hand digit is not really all that much more work. Once you become fully used to it, you will find it quicker and simpler than the old way.

Let's review for a moment what we mean by left-hand and right-hand digits in multiplication. Try to "see" the *left-hand* (tens) digit of

$$\begin{array}{r} 4 \\ \times\ \underline{2} \end{array}$$

If you fully mastered the last chapter, you answered, almost without a second thought, "zero."

What is the right-hand (units) digit?

If the digit 8 sprang into your mind with little or no effort, you are already well on the way to accelerating your multiplication with the no-carry method. If you had to stop and think, however—as most of us do at this point—then that is exactly what this chapter is for.

Your first exercise is to go through the following digit pairs with the object of "seeing" only the left-hand (tens) digit—the one we have been describing as "is in the 20's, 70's," etc. See and think, as well as you can, 4 × 4 as "1."

The first time you go through these, it might be wise not to try for speed. The first job is to begin training the habit of recognizing the left-hand digit automatically. Just as important, you should build the habit of thinking "zero" when there is no real left-hand digit (that is, when the full product is less than ten) because this is so important to accuracy in multiplying longer numbers.

Remember to see 4 × 9 as "3"—*not* "the left-hand digit of 4 × 9 is 3, because 4 × 9 is 36 and 36 is in the 30's"—just as you see *u* and *p* as "up."

Go slowly and carefully this first time, training your mind to see only the answer. Left-hand (tens) digits only:

$$\begin{array}{cccccc}
7 & 4 & 1 & 6 & 8 & 4 \\
\underline{4} & \underline{7} & \underline{2} & \underline{4} & \underline{5} & \underline{3} \\
 & & & & & \\
3 & 8 & 2 & 9 & 1 & 2 \\
\underline{6} & \underline{2} & \underline{5} & \underline{3} & \underline{4} & \underline{7} \\
 & & & & & \\
5 & 9 & 1 & 5 & 6 & 4 \\
\underline{3} & \underline{5} & \underline{8} & \underline{8} & \underline{6} & \underline{1} \\
 & & & & & \\
6 & 3 & 8 & 9 & 5 & 7 \\
\underline{7} & \underline{2} & \underline{9} & \underline{7} & \underline{5} & \underline{2}
\end{array}$$

That is enough for the first dose. You will go through every possible digit combination before you are through, but doing them all at once might become tedious.

Compare, if you will, the study of speed mathematics to learning any new skill. There is a specific objective in mind, of course—in this case, to solve problems more rapidly and easily. But there is also a helpful secondary objective: becoming fascinated with the process of *doing* and excited about your mastery of the technique. Just as a craftsman enjoys the actual process of making a perfect joint in a woodworking project because it is satisfying to do something skillfully, so can you become fascinated with the dispatch and accuracy of your working of a sample problem in a new way.

When we use them in business, numbers always stand for something. When we practice with them, however, they become an impersonal sort of puzzle. Look on them as a crossword puzzle, or a chess problem, or a brain-teaser. Just as satisfying as these and far more rewarding—because your growing skill at this type of puzzle will pay you solid dividends for the rest of your life.

Now, carefully rather than hastily this first time, continue working at the habit of seeing only the left-hand (tens) digits of these combinations:

8	7	4	3	6	4
7	3	9	3	1	6

3	1	6	5	3	2
4	1	5	9	5	8

2	7	1	8	8	6
9	8	6	4	1	9

5	4	3	5	2	2
2	5	8	7	2	1

By now you should find the habit beginning to take hold. Once the proper response starts becoming a habit, you can go back over the examples with the objective of speeding up your reaction time.

Make very sure at this point, though, that you work at giving your response in the right fashion, rather than giving a fast but improper one. Going reasonably slowly now will contribute to greater speed in the future.

Now finish your practice on left-hand (tens) digits with the rest of the basic digit combinations:

1	9	3	5	1	3
9	2	9	6	7	1

4	9	8	9	6	3
2	8	6	9	8	7

6	5	4	8	1	4
3	4	4	3	3	8

7	2	7	6	9	5
6	3	5	2	1	1

7	2	9	8	9	2
1	4	4	8	6	6

1	7	7	9	7	4
5	7	8	8	9	9

Stop and take stock of your technique now. Do you find that you are looking only for the left-hand digit as you glance at each pair? Have you schooled yourself to give only the answer? Are you *always* thinking "zero" when the product of the two digits is less than ten?

If not, make a point of going back over the combinations from time to time, working specifically to develop this habit. If you feel that you are making the proper responses a routine, then your next step should be to develop speed. Time yourself in completing the tables, and make a note of how long it took. Next time, see if you can shave a few seconds off the last record.

So far, you have worked at accuracy and speed in seeing

Right-Hand Digits

only the left-hand, or tens, digit of each product. This is only half the story. The other half is to do precisely the same thing for what the product "ends in."

Glance at this example:

$$\begin{array}{r} 6 \\ \times\ \underline{7} \end{array}$$

What is the left-hand digit?

What is the right-hand digit?

There would be little point to repeating all the tables again just for the right-hand digit practice. Instead, use the same tables on the last few pages.

Keep in mind the important practice points mentioned in connection with left-hand digits. Go slowly the first time, consciously making an effort to "see" only the right-hand digit, rather than the problem. You may find it helpful to say the answer to yourself; if you do, be very careful not to say the problem.

After you have gone over the tables just a few times, you should begin to find yourself simply reading the answers—just as you read these words or phrases rather than the letters.

If you need proof of this, stop right now and try to recall whether there were any *f*'s in the paragraph above. In all likelihood, you haven't the vaguest idea. You undoubtedly read the first word "after" without even noticing the *f* in it. In the same way, you can approach this "end-result-only" ability with digit combinations.

Go back to the tables and do your first right-hand digit practice now.

Work at the tables conscientiously, but I would suggest that you alternate practicing the digit combinations with some of the other practice to come. Avoid the stale, "overtrained" reaction of too much consecutive time spent at only one part of the whole.

Two-Digit Practice

The whole reason you practice the basic multiplication

table with left-hand and right-hand digits is so you can multiply from left to right without carrying. Picking up the right-hand digit from one product and adding it to the left-hand digit of the product to the right is the secret that eliminates carrying altogether.

You do have to keep one digit in your mind for a moment, but this is considerably simpler than juggling three (and sometimes four) digits in traditional right-to-left multiplication.

Refresh your memory with this example:

$$\begin{array}{r} 7\ 8 \\ \times\ 4 \\ \hline \end{array}$$

See if you can anticipate each step of this review:

Step one: 7 × 4 is in the 20's:

$$2$$

Step two: 7 × 4 ends in 8. 8 × 4 is in the 30's. 3 minus 2 (complement of 8) is 1, and record the ten:

$$\underline{2}\ 1$$

Step three: 8 × 4 ends in 2:

$$\underline{2}\ 1\ 2, \text{ or } 312$$

The step two above is as complicated as no-carry multiplication can ever get. You have to remember the 8 while getting the 3. If you have learned to read answers, you would think only "8, 3, 1, record." The same point in schoolbook multiplication would involve these thoughts: "Carry the 3 from 32. 4 × 7 is 28. Add the carried 3 to 8, which makes it 11. Put down 1 and carry 1 to the 20. Put down 3."

This review is to encourage you to spend some of your practice time on the two-digit tables that follow. It would be impractical to include every possible combination (there are just under a thousand of them), but you will find a good spread of every type.

The first time you do this section, work slowly and evenly, disciplining yourself to think along the lines we have covered:

Read only the answer to each digit combination.

Work from left to right.

Think an initial zero if this is the left-hand digit of the first product.

Add the center digits of the answer with a complement if it goes over ten, and mentally record the ten by underlining the imaginary digit to the left in the answer.

Say aloud the answers to these problems:

7 4	6 8	3 8	9 1	5 9	6 4
× 1	× 4	× 2	× 2	× 5	× 6
3 8	5 7	8 7	4 3	3 6	5 3
× 9	× 8	× 2	× 7	× 3	× 4
2 2	4 1	5 9	5 8	2 7	1 8
× 8	× 6	× 9	× 6	× 4	× 2
8 6	5 4	3 5	5 2	2 2	6 8
× 7	× 2	× 2	× 9	× 3	× 5
9 0	5 1	3 4	9 8	6 7	2 0
× 9	× 2	× 3	× 6	× 8	× 1

The most important thing you undoubtedly noticed is that your ease with these problems is based very directly on your ability to read automatically the left- and right-hand digits of the products of each combination. If they pop into your mind without thought—as they will after surprisingly little practice—then expanding your practice to two-digit examples is almost painless. But if you have to stop and think hard to get each digit, then you will find this section much harder and slower than it should be.

If you experienced trouble in "reading" the left- and right-hand digits to make these problems easy, go back and review your single-digit tables once or twice before going on. Each time you do them, the answer will come a little more automatically.

Now read from left to right the answers to these problems. Make sure you are building the right habits as you do so. Make it a point to use the proper technique:

9 4	8 7	6 2	4 9	9 7	7 7
× 6	× 9	× 4	× 2	× 8	× 5
1 7	5 8	4 7	7 3	3 0	9 3
× 3	× 6	× 4	× 8	× 5	× 1
6 7	3 8	2 4	9 1	6 5	1 2
× 2	× 8	× 3	× 7	× 4	× 6
2 8	1 4	7 2	8 5	4 6	9 8
× 3	× 8	× 4	× 9	× 6	× 2
5 5	6 4	4 1	5 8	7 9	4 7
× 8	× 2	× 9	× 7	× 5	× 4

The final practice table of this chapter follows. You have already practiced all the essentials. If you can handle two-digit tables with snap and decisiveness, then you can keep on doing step two through twenty-digit problems. You already know how to line up your columns for multipliers of two digits or more, and you know how to add more effectively and quickly than ever before. The final section asks you to say aloud, from left to right, the answers to a variety of multiplications with single-digit multipliers but differing numbers of digits in the numbers multiplied.

Just as you did with both the one-digit and two-digit tables, work slowly and carefully the first time over this varied practice group. Get the foundation of proper habits firmly established. Say your zero first digits where they are required, think an underline to the left when you use a complement, and do your very best to read only the answer to each combination—not the combination itself.

For longer problems you may wish to write down your answer. Just put your pad under the problem and jot down the answer from left to right, as it develops naturally in your mind.

Spend several minutes at this:

7 × 9	23 × 9	456 × 7	3 × 4	782 × 2
81 × 3	6 × 9	22 × 8	6843 × 7	91 × 6
38 × 3	87 × 4	543 × 5	5 × 2	453 × 8
6 × 3	817 × 7	729 × 2	7 × 1	9478 × 6
974 × 3	81 × 9	285365 × 8		26 × 4
41 × 4	684 × 7	3 × 2	82 × 5	913 × 3

The mixing of problems with one, two, three, and more digits in this section was intentional. This is the way problems are presented to us in business. They do not ordinarily come neatly packaged in orderly rows of similar problems. Your mastery of each method and technique always has to become adaptable as well as proficient.

8
SHORT-HAND DIVISION

SO FAR, we have covered three out of the four basic arithmetical computations.

In adding, we learned to use complements for the tougher combinations—those that would add over ten if we ever added over ten—and to record tens in such a way that the answer forms naturally in our mind, just as, on the modern abacus, the answer forms naturally on the board.

In subtracting, we learned never to subtract a larger digit from a smaller, and to avoid that crude and precarious method of "borrowing," so that again the answer forms itself easily and naturally in the mind or on the paper.

In multiplying, we have torn apart the multiplication table so that we use only half of it at a time. This enables us to discard the idea of "carrying," and furthermore produces the answer from left to right. When we have to record tens in preceding digits in our answer, we adopt a simple and effective method that—again—gives us a natural development of the answer.

Now, what about dividing?

There is no *single* secret for speed division comparable to the secrets of complements or no-carry multiplication. But by leaning on *both* complements and no-carry multiplication, we can build a streamlined technique for division that, in its total effect, can save you as much time and effort in this field as the single secrets can in theirs.

In order to get our ground firmly established, let us look at a sample problem in division and work it in the traditional long-division way:

```
              5 2 4
386 / 2 0 2 2 6 4
      1 9 3 0
      -------
        9 2 6
        7 7 2
        -----
        1 5 4 4
        1 5 4 4
        =======
```

This is a fairly simple problem. It has no remainder. Everything comes out even. Yet a great deal of pencil work was involved. It *looks* complicated.

Just for comparison, although the figures will be meaningless to you at the moment, let us show what the same problem would look like in the shorthand method you will learn in this chapter:

```
              5 2 3
                  -
386 / 2 0 2 2 6 4
      1̸ 1̸ 9
         2̸ 5 4
         1̸ 4̸ 9̸ 6
```

Certain elements of these numbers should be familiar to you—the underline and the slashes. The shorthand method will rely on your confident handling of complement subtraction and no-carry multiplication.

The two hardest parts of traditional long division, you will undoubtedly agree, are (first) determining at a glance the next digit of the answer and (second) going through the complex pencil work of verifying that digit and finding the remainder into which you divide in order to determine the next digit of the answer.

This chapter will offer a simpler way of doing each of

these processes. Before we go into them, however, consider a few basic facts about the process of division.

Continuous Approximation

Long division, by which we mean division by a number of several digits, is really a progressive estimate that gets more accurate as we finish more of it.

In this sense, division is radically different from adding, subtracting, or multiplying. It is the only one of the four processes that we were taught to do from left to right, in the natural way. Since this is true, division is already self-estimating, just as the new methods for doing the other three processes are.

The familiar process we call long division, incidentally, seems to be a special crutch developed only in England and America, which, because every single step is spelled out (and written down in detail), no rational person in school can fail to learn to handle. It is certainly accurate and easy enough, but it is also infernally slow and cumbersome.

For another comparison, look at the division we just examined, next to the same problem solved in the European short-hand method:

```
        American                     European

               5 2 4                        5 2 4
3 8 6 / 2 0 2 2 6 4          3 8 6 / 2 0 2 2 6 4
        1 9 3 0                             9 2 6
        -------                             1 5 4 4
          9 2 6
          7 7 2
          -----
          1 5 4 4
          1 5 4 4
          =======
```

If you have never before confronted this European (in England it is called "Continental") method, you may feel some awe of European education when you learn that the difference is simply this: the multiplying and subtracting are done entirely in the head. They are never written down at all. The two lines of working figures you see under the problem are merely the *results* of each subtraction.

Difficult? To us, yes. To a French or German schoolboy it is something he is expected to learn; and learn it he does, or flunks out and spends the rest of his life hoeing potatoes. But to those of us trained in the "dot-every-*i*, put-down-every-digit" methods of American arithmetic, it is rather difficult to learn late in life.

What you will discover before this chapter is over, however, is that applying your new mastery of simplified left-to-right multiplication and subtraction will make it not only possible to divide in a way similar to the European method, but actually easier than it was in the standard long-division way.

Before we get into this subject, we will first explore a method for rapid answer-producing that removes the first major stumbling block to quick and easy dividing.

Automatic Division

The point at which most of us hesitate longest in working our way through any long division is deciding on the next digit of the answer.

Consider this example:

$$87 \overline{)4263}$$

The first step is to divide 87 into 42 or, since this "won't go," into 426.

Almost all of us, no matter how good our number sense is otherwise, lack any sort of genuine feel for such an answer. We are not dividing by 8, but by 87. Think back, and you will probably find that you often try two or three "trial answers" in your mind before deciding on one to put down.

Here is a simple trick to overcome this difficulty—a trick that automatically delivers to you the next digit of your answer no matter how complicated the divider is. It makes dividing by 34,968 as simple (at this point) as dividing by 4.

The trick is this: Do not divide by the divider. Divide only by its first digit, *raised by one*. Do not divide into the number divided. Divide only into its first digit (if that digit is larger than the divider digit) or into its first two digits (if the first digit is smaller than the divider digit).

This technique, by the way, is also adapted with minor variations from modern soroban theory. It is considered as basic to speed and ease on the abacus as is the use of complements for adding and subtracting.

In the example above, you do not try dividing 87 into 4263. Instead, divide 9 into 42. This is much, much easier. You should "see" the answer 4 at a glance.

The reason this works is that 87 is somewhere between 80 and 90, but for simplicity we consider it to be 90. A little over half the time, this first digit will be correct. Less than half the time, it will need revision—but the revision will be automatic and quick, just as it is on the abacus.

Try this technique on these examples:

$$47\,/\overline{268} \qquad 65\,/\overline{513} \qquad 28\,/\overline{136}$$

On these three problems, our automatic division works like this:

5 (instead of 47) into 26 (instead of 268) is 5. This is the correct first digit of the answer.

7 (instead of 65) into 51 (instead of 513) is 7. Right.

3 (instead of 28) into 13 (instead of 136) is 4. This also checks out.

Caution: Note with special care that using this trick to "see" each successive digit of your answer does *not* alter the position of each answer digit. In the first example, you put the answer digit 5 over the 8, not over the 6. You still follow your classical rule for placing your answer: start as many digits over in the number divided as there are digits in your divider—plus one if you start by dividing into two digits instead of one:

$$\begin{array}{r} 1 \\ 76\,/\overline{9863} \end{array} \qquad \text{but} \qquad \begin{array}{r} 2 \\ 76\,/\overline{1732} \end{array}$$

In the first case, we "see" 8 into 9 and put down the answer digit 1 two places to the right because there are two digits in the divider. In the second case, we "see" 8 into 17 and put down the answer digit 2 three places to the right because there are two digits in the divider *and* we started the division into two digits of the number divided.

Now get the idea firmly in hand by trying these:

5 6 / 2 1 6   9 4 / 1 6 8   3 7 / 8 3 6

Remember that we are not yet finishing these divisions. At the moment, we are concerned only with developing this rapid and foolproof way to produce automatically each digit of the answer without hesitation.

Check your reactions to the above three examples. Did you see the first as 6 into 21, and put down 3 over the 6? Did the second become 10 instead of 9—indicating that the answer digit 1 goes over the 8? When you got to the third, did you "see" 4 into 8 as 2, and put it over the 3? If any of your answer digits got misplaced, review the general rule once more:

If your first division is into a single digit (4 into 9), the first answer digit appears as many places to the right over the number divided as there are digits in the divider.

If your first division is into two digits (4 into 23), the first answer digit moves one more place to the right.

The principle of finding each digit of the answer by dividing with only the first digit of the divider, raised by one, works with problems of any length. Experiment only on the examples provided, however, until we come to automatic revision.

Go through the following problem on your pad. Find the two digits of the answer by dividing with the first digit only (plus one) of the divider:

8 7 6 / 5 7 1 4 0

Work this out completely in your traditional handling of long division, applying to it at the moment only the new automatic digit-finder.

The final answer is 65. The first digit is produced by dividing 9 (not 876) into 57 (not 5714) and putting the resulting 6 over the 4. When you multiply out and subtract, you divide into the remainder 4580 for the second digit. 876 might make you hesitate between 5 and 6 for the second answer digit, but 9 into 45 can only be 5. We have produced two digits of the answer by simple inspection. For now, we will ignore the fractional remainder.

Now we will go on to the special aid that makes this technique useful on any problem at all, not merely on carefully selected examples.

Automatic Revision

Consider this case:

$$876 \,/\overline{4380}$$

Start with the trick of dividing 9 into 43, instead of 876 into 4380. 9 will go into 43 no more than 4 times. Yet if you multiply out the divider by 4 and subtract, you find a remainder of 876. This is the divider itself. The answer to this problem is 5, not 4. Then it comes out even.

What is wrong? Nothing, really. We said earlier that division is really a continuous approximation from left to right. The digit of the answer we first put down is an approximation that may need revising before we finish.

On the abacus, each trial digit is produced by dividing with the first digit of the divider but without raising it first. Revision is frequently necessary, just as it is in this system. But revision on the abacus is always to *reduce* the trial digit by one (sometimes two), adding in this revision factor to the trial remainder. In our system of using a digit that is one higher than the first digit of the divider, the only way we ever have to revise is *upward*. As you will see when the technique develops fully, this is easier and more automatic with pencil and paper.

Since you are in effect dividing by a number larger than your real divider, you could not possibly try too large an answer digit. It is child's play to revise your answer upward in our system, but it would be quite difficult to revise it downward.

You have learned in no-carry multiplication how to increase the value of a digit by one without rewriting it. You simply underline it. The answer to the example above, when finished, would look like $\underline{4}$. You read it as 5.

All of this will be drawn together as we assemble the various parts of the complete division system. For the moment, remember only that you speed up your division by "seeing" the answer to 9 / 43 rather than trying to work out an answer to 876 / 4380.

Try this part of the technique once more. Do not bother to complete these examples. Just practice "seeing" the first answer digit by dividing with the first digit only of the divider, raised by one:

230 / 6843   78 / 149

84375 / 684219

Now we will combine everything we know about multiplication and subtraction, both of which are continuously involved, with this simplified digit-finding technique, to make the complete shorthand division method both easier and faster than the cumbersome method of long division.

Shorthand Division

We began the explanation of no-carry multiplication by taking apart a sample problem and seeing how the answer develops. Let us do the same thing with a sample division:

```
      6 3
7 / 4 4 1
    4 2
    ---
      2 1
      2 1
```

We have called the process of division "continuous approximation." The first approximation you got in the above problem was really 60, not 6: 7 goes into 441 something more than 60 times. You know this because there is obviously another answer digit to come.

We get the second answer digit by finding out first how much of the 441 is left after subtracting from it exactly 60 7's. In long division, we multiply the answer digit by the divider and put this product under the portion of the number divided that produced the digit.

That product here is 420. We do not ordinarily bother with the 0, any more than we bothered with the 0 in 60, since careful placement of each digit takes them into account.

By subtracting 42 from 44 and then "bringing down" the remaining 1 in the number divided, we find that there is 21 left over. Actually, we really subtracted 420 from 441. The "bringing down" completes that process. It would be helpful to inspect two expressions of this situation:

$$\begin{array}{r} 6\ 0 \\ 7\ /\overline{4\ 4\ 1} \\ 4\ 2\ 0 \\ \hline 2\ 1 \end{array} \qquad \text{is the same as} \qquad \begin{array}{r} 6\ 0 \\ \times\ 7 \\ \hline 4\ 2\ 0 \\ +\ 2\ 1 \\ \hline 4\ 4\ 1 \end{array}$$

Dividing now the 7 into the remainder, we find that it will go exactly 3 times. In long division, we verify this by multiplying 7×3 and subtracting it from the remainder, getting a final remainder of 0.

Again, try to feel the identity of these two expressions of the current situation:

$$\begin{array}{lrcr} 7 & /\overline{4\ 2\ 0} & = & 6\ 0 \\ 7 & /\overline{2\ 1} & = & 3 \\ \hline & 4\ 4\ 1 & & 6\ 3 \end{array} \qquad \text{is the same as} \qquad \begin{array}{rcrcr} 7 & \times & 6\ 0 & = & 4\ 2\ 0 \\ 7 & \times & 3 & = & 2\ 1 \\ \hline & & 6\ 3 & & 4\ 4\ 1 \end{array}$$

Now we will accomplish the same result with a fraction of the pencil work involved in long division.

The two most tedious parts of long division are (1) multiplying the answer digit by the divider and writing it down as you go, and (2) subtracting this product from the part of the number divided involved, in order to establish the remainder so far.

The European system, you recall, involves doing these two steps in your head. You write down only the final result of each subtraction. But this involves handling several digits at once in your head—contrary to the best approach to speed mathematics.

Since you know how to multiply from left to right, digit by digit, and also how to subtract from left to right, digit by digit—*without carrying or borrowing*—you can combine the two and accomplish the European result *without* ever handling more than one digit at a time.

We will use that same problem as the first model:

```
        6
 7 / 4 4 1
```

The following process, remember, is multiplication and subtraction done in one-two order, one digit at a time:

One: 7 × 6 is in the 40's, and 4 from 4 is 0:

```
        6
 7 / 4 4 1
     -
```

We do not bother to write the 0. As you become accustomed to this system, you will not even bother to make any mark at all for this result. The *lack* of a digit there shows you that the result was 0.

Two: 7 × 6 ends in 2, and 2 from 4 is 2:

```
        6
 7 / 4 4 1
     - 2
```

Traditional long division would now require you to rewrite the next digit of the number divided—1—beside the 2. You do not need to do this. You can bring it down mentally and see that the remainder is now 21, by reading the problem like this:

```
        6
 7 / 4 4 1
     - 2 ↙
```

The next digit of the answer is 3, and you know this is right merely by inspection. Just to get the technique thoroughly established, however, we will verify it as you would in a more complicated problem:

One: 7 × 3 is in the 20's, and 2 from 2 is 0.

Two: 7 × 3 ends in 1, and 1 from 1 is 0.

Compare the work you have now finished with the same problem spelled out in long division:

```
       6 3              6 3
  7 / 4 4 1        7 / 4 4 1
      4 2              - 2
      ---
        2 1              - -
        2 1
```

Get out your pad and pencil and actively follow each step in this demonstration:

```
        7
  8 / 5 9 3 6
```

Although we have not mentioned it before, you naturally divide by any single digit without raising it in value by 1. If this divider were 84, we would divide by 9 because 84 is somewhere between 8 and 9. But 8 is obviously nothing but 8.

The first answer digit, by inspection, is 7. Now we multiply and subtract digit by digit:

One: 8 × 7 is in the 50's, and 5 from 5 is 0.

Two: 8 × 7 ends in 6, and 6 from 9 is 3.

```
        7 4
  8 / 5 9 3 6
      - 3
```

The remainder so far is 336. In order to produce the next answer digit, we mentally bring down the 3 and divide 8 into 33. We put down the answer digit 4. Now we verify and produce the remainder:

One: 8 × 4 is in the 30's, and 3 from 3 is 0.

Two: 8 × 4 ends in 2, and 2 from 3 is 1.

```
       7 4 2
8 / 5 9 3 6
     - 3 0
      - 1
```

The remainder at this point is 16. In the illustration above, we have already mentally brought down the 6 and put down the next answer digit, 2. Is there any remainder?

One: 8×2 is in the 10's, and 1 from 1 is 0.

Two: 8×2 ends in 6, and 6 from 6 is 0.

The problem comes out even. A little later on, we shall get into fractional and decimal remainders.

Now try one entirely on your own. It will be an easy one, to get the technique firmly bedded in your habits before going on to more complicated problems. Do this one on your pad:

```
6 / 4 5 6
```

After you have finished, check your working against this step-by-step explanation:

First digit: 7. One: 6×7 is in the 40's, and 4 from 4 is 0. Two: 6×7 ends in 2, and 2 from 5 is 3. Remainder (by mentally bringing down the 6), 36.

Second digit: 6. One: 6×6 is in the 30's, and 3 from 3 is 0. Two: 6×6 ends in 6, and 6 from 6 is 0.

Automatic "Borrowing"

The demonstrations so far have been chosen for simplicity. They are simple both because you are dividing by single digits and because there is no canceling ("borrowing") involved in the subtraction.

Consider this case:

```
9 / 2 2 2 3
```

This problem will involve canceling. Yet because you have learned to use canceling in the answer instead of "borrowing" in the larger number, you will find it no trick at all to adapt what you already know to the smooth and efficient working of this kind of division.

The first answer digit, by inspection, is 2. We get the remainder so far in our usual way:

One: 9 × 2 is in the 10's, and 1 from 2 is 1.

Two: 9 × 2 ends in 8, and 8 from 2 is —

STOP! Larger from smaller. Do not subtract. Add the complement of 8 (2) to 2 and slash left:

```
        2
9 / 2 2 2 3
    1̸ 4
```

This answer should look perfectly normal after your work with left-to-right subtraction. It is simply 4. The (slashed) 1 has been reduced by the slash by one in value, to 0.

Mentally bringing down the next 2, you "see" the answer of 9 into 42 as 4 and put this down as the second answer digit. Now for the remainder:

One: 9 × 4 is in the 30's, and 3 from 4 is 1.

Two: 9 × 4 ends in 6, and 6 from 2 is—larger from smaller. Do not subtract. The complement of 6 (4) plus 2 is 6, and slash:

```
        2 4
9 / 2 2 2 3
    1̸ 4
      1̸ 6
```

See if you can finish this problem yourself. Mentally bring down the proper digit and see the answer. After you have worked it out, check against this explanation:

The next answer digit is 7—9 into 63.

One: 9 × 7 is in the 60's, and 6 from 6 is 0.

Two: 9 × 7 ends in 3, and 3 from 3 is 0.

Try one example that involves "borrowing" before going

on. Use your pad and do this problem just as we did the one above:

```
8 / 5 4 2 4
```

Do this with your pad and pencil before checking against the working figures below. A full understanding of the steps in shorthand division is essential before we get into longer problems. Once you have gone through the routine several times, you will find that you can handle any division with confidence.

Check your work and your answer with this finished problem. Here is how it should look:

```
      6 7 8
8 / 5 4 2 4
    1̸ 6
      1̸ 6
```

That is all there is to it. No single part of this process is complicated. It is all based on techniques you have already mastered in earlier parts of this book. But the combination of them is new. If you have any trouble assembling the parts into a smooth-working whole, then go back and re-check the weak parts.

Make very sure you have everything so far down pat, because we are now going to add the third and final element in shorthand division that makes it just a bit more complex. You have already learned to handle this step in multiplication, but the mental processes will have to stretch one more notch when you apply it to division.

Stop now and make sure you are ready for the next step by doing these two problems on your pad:

```
4 / 1 8 0          8 / 6 2 3 2
```

When you have done these, compare your results with these models:

```
      4 5                7 7 9
4 / 1 8 0          8 / 6 2 3 2
      2                1̸ 6
                         1̸ 7
```

Now let us pick up the final technique from multiplying that enables you to handle shorthand division with dividers of any length.

No-Carry in Division

So far, you have been dividing by only one digit. In such divisions, you would ordinarily do most or all of these steps entirely in your head anyway and not worry about putting down the remainders as you went along. It is really short division, and we have started with this only to get the general technique firmly established.

When you divide by numbers of two or more digits—by 653, for instance—you will add to your simultaneous left-to-right multiplication and subtraction the efficient and handy no-carry system.

This is the point where the European shorthand method becomes really difficult for most of us. When dividing by a number of two or more digits, the European system requires you to multiply (including carrying) and subtract (including borrowing), all in your head. This can involve juggling as many as six digits all at once in your mind. With our left-to-right methods, however, we can do everything digit by digit.

If you understand thoroughly both the no-carry multiplication method and the division method covered so far, then you could probably work out the entire method yourself without further help. Since it is using no-carry multiplication within a new framework, however, we will go into the entire process step by step.

Recall, as we get into this, the technique of dividing by only the first digit of the divider—raised by one:

$$3\ 8\ /\overline{1\ 3\ 5\ 2\ 8}$$

Our first answer digit we "see" by dividing 4 into 13, and putting 3 over the 5. Remember that the first answer digit starts as many places over the number divided as there are digits in the divider—plus one if you start dividing into two digits instead of one.

Now we develop the remainder:

One: 3 × 3 is in the zeros, and 0 from 1 is 1:

```
          3
3 8 / 1 3 5 2 8
      1
```

Two: 3 × 3 ends in 9. 8 × 3 is in the 20's. 9 + 2 is (complement of 9 from 2) 1, and record. Now you have *two* things to do: subtract 1 from 3, and record the ten. Since we are subtracting while we multiply, this ten obviously gets *subtracted.* How? Just by canceling; slash left:

```
          3
3 8 / 1 3 5 2 8
      1̸ 2
```

Three: 8 × 3 ends in 4, and 4 from 5 is 1:

```
          3
3 8 / 1 3 5 2 8
      1̸ 2 1
```

There was a curve hidden in point two of that example, but it seemed best to slide it in quietly. It is a new application of recording by slashing, because the only digits we jot down are the results of the subtraction and the final effect of a recorded ten from the multiplication is obviously to *reduce* the preceding digit in the answer to the subtraction by 1.

So one of the side-rules of speed division grows from this: when your no-carry multiplication involves a complement (and therefore a recorded ten), slash the digit to the left in your working figures.

Our remainder so far is 2,128. 38 will go into 13,528 300 times, with 2,128 left over. If the size of these figures jars you, inspect the work so far and think back to the identity-expressions at the start of this section.

Note that we have two digits in the answer to the subtraction. We still bring down the next digit (mentally), so the next division is 38 into 212. We "see" it as 4 into 21, and put down 5. Now we develop the remainder:

One: 3 × 5 is in the 10's, and 1 from 2 is 1:

```
          3 5
38 / 1 3 5 2 8
     1̸ 2 1
       1
```

Two: 3 × 5 ends in 5. 8 × 5 is in the 40's. 5 plus 4 is 9, and 9 from 1 is—larger from smaller. Add the complement of 9 to 1, and put down 2. Cancel by slashing to the left:

```
          3 5
38 / 1 3 5 2 8
     1̸ 2 1
       1̸ 2
```

This brings up an interesting point. You have *two* occasions to slash to the left when doing shorthand division: when a complement is used in no-carry multiplication, and when a complement is used in subtraction. Both involve the use of a complement, and both result in a slash to the left. One special result of this will develop later.

Three: 8 × 5 ends in 0, and 0 from 2 is 2:

```
          3 5
38 / 1 3 5 2 8
     1̸ 2 1
       1̸ 2 2
```

The remainder at this point is 228. This is the excess after subtracting 38 × 350 from 13528.

Before going on to the final digit of this answer, and determining the remainder (if any), recall our earlier comments about revised digits. Dividing by only the first digit of the divider is the quickest and easiest way to produce the next digit of the answer, but sometimes it will need revising. This is a price paid happily by operators of the high-speed abacus, because it saves more time in producing each digit than it costs in revision.

The final digit of this answer will demonstrate such a case.

See the next digit of the answer as 4 into 22—5. Now let us work out the remainder (if any) and see what happens.

One: 3 × 5 is in the 10's, and 1 from 2 is 1:

```
           3 5 5
3 8 / 1 3 5 2 8
      1̸ 2 1
        1̸ 2 2
            1
```

Two: 3 × 5 ends in 5. 8 × 5 is in the 40's. 5 plus 4 is 9, and 9 from 2 is—complement of 9 plus 2, and slash:

```
           3 5 5
3 8 / 1 3 5 2 8
      1̸ 2 1
        1̸ 2 2
          1̸ 3
```

Three: 8 × 5 ends in 0, and 0 from 8 is 8:

```
           3 5 5
3 8 / 1 3 5 2 8
      1̸ 2 1
        1̸ 2 2
          1̸ 3 8
```

The remainder is the same as the divider. So our final answer digit must be raised by 1, and the problem will come out even. As we pointed out before, you raise the digit by underlining it, so the final answer is

3 5 $\underline{5}$

which you read as 356.

You were promised that digit-revision would be automatic. It is. Any digit in your answer may need revising, even the first. But the time to do so is signaled to you automatically, so you do not need to watch especially for such events.

This problem demonstrates why:

```
6 3 / 5 4 7 4 7
```

We "see" the first answer digit as the result of 7 into 54, which can be only 7. Now (use your pencil and pad to help build the technique) we find the first remainder:

One: 6 × 7 is in the 40's, and 4 from 5 is 1:

```
          7
63 / 5 4 7 4 7
     1
```

Two: 6 × 7 ends in 2. 3 × 7 is in the 20's. 2 plus 2 is 4, and 4 from 4 is 0. Where a zero digit appears in the *middle* of a subtraction answer, as it does here, it is wise to put it down to avoid possible confusion:

```
          7
63 / 5 4 7 4 7
     1 0
```

Three: 3 × 7 ends in 1, and 1 from 7 is 6:

```
          7
63 / 5 4 7 4 7
     1 0 6
```

Perhaps you have noticed that your remainder in this case is larger than your divider. Something is wrong, and what is wrong is that your answer digit needs raising by one.

You do not have to be especially alert to this situation, however. If you didn't notice at this point, you could not help but notice as soon as you tried to get the next answer digit. You mentally bring down the 4 and divide 63 into 1064—seeing it as 7 into 106.

Such an answer digit would be over ten. There is no such digit. This, in case you missed the signal that developed when your first remainder was larger than the divider, is the STOP signal that warns you to revise your answer digit.

This is automatic division. To raise the answer digit, you merely underline it. To adjust the remainder, you merely *subtract the divider* and put down the new remainder before going on:

```
          7
          -
63 / 5 4 7 4 7
     1 0 6
     1 4 3   (the 1 is struck through)
```

The underlined 7 is, of course, now 8. The 43 is the

answer we get after subtracting 63 (the divider) from 106 (the too large remainder). Now we are ready to continue, with everything adjusted and correct so far.

Finish this problem yourself on your pad. It does come out even, though one other answer digit will need revision. Finishing this problem will involve just about every technique in shorthand division covered so far.

Do it now.

If your final answer did not come out to an even $\underline{7}6\underline{8}$, which you read or rewrite as 869, check the appearance of your working figures against this model:

$$
\begin{array}{rl}
 & \underline{7}\;6\;\underline{8} \\
6\;3\;/ & \overline{5\;4\;7\;4\;7} \\
 & 1\;0\;6 \\
 & \not{1}\;4\;3 \\
 & \not{1}\;\not{0}\;6 \\
 & \not{1}\;6\;3
\end{array}
$$

This section, in its joining together of many different techniques from earlier parts of the book into one effective but apparantly complex whole, has been perhaps the most difficult chapter to understand in one reading.

Sit back for a moment and let some of what you have done sink into your mind. Don't be discouraged if it takes several readings to understand fully what has been going on. It involves quite a new way of looking at numbers, a way really simpler than the traditional ways because much of it has been adapted from the simplest and highest-speed arithmetical system known—the modern Japanese abacus—but until you get used to it it does take some special lip-biting.

Division is the most complex of all our basic operations in arithmetic. There is simply no help for this; it is the nature of the beast.

What we have done so far is to reduce the process to the simplest series of easy steps that can possibly work. You never hold more than a digit or two in your mind at any one point; you work from left to right; you never have to carry as you multiply; and you never have to "borrow" as you subtract.

The seeming complexity at this point is inherent in the function itself. If you had never learned long division, and if somebody sat down to explain it to you, it would seem much more complex. There are many separate processes to be done, and if full accuracy is required every process must be done in full.

Except for one really minor special case, you now know everything you need to know for this rapid way to divide. Ease and speed will come with practice, which the next chapter will help to provide.

The one special case involves slashing a number not once, but twice. We hinted at this possibility when we pointed out that you slash left when you use a complement in multiplying, and you also slash left when you use a complement in subtracting. It does not come up very often, but it does come up and it is very easy to handle.

Here is the sort of problem in which you will find this necessity:

$$74 \,/\, \overline{5\ 0\ 5\ 4\ 2}$$

Let's begin this problem step by step, to drive the method deeper into your mind.

8 (not 74) into 50 (not 505) gives 6 as the first answer digit. Put it down, and determine the remainder:

One: 7 × 6 is in the 40's, and 4 from 5 is 1.

Two: 7 × 6 ends in 2. 4 × 6 is in the 20's. 2 plus 2 is 4, and 4 from 0 is (complement) 6 and slash.

Three: 4 × 6 ends in 4, and 4 from 5 is 1.

Our example now looks like this:

$$\begin{array}{rl} & 6 \\ 74\ / & \overline{5\ 0\ 5\ 4\ 2} \\ & \not{1}\ 6\ 1 \end{array}$$

Inspection shows us that the next answer digit (8 into 61) is 7. We put it down and work out the remainder:

One: 7 × 7 is in the 40's, and 4 from 6 is 2:

$$\begin{array}{rl} & 6\ 7 \\ 74\ / & \overline{5\ 0\ 5\ 4\ 2} \\ & \not{1}\ 6\ 1 \\ & 2 \end{array}$$

Two: 7 × 7 ends in 9. 4 × 7 is in the 20's. 9 plus 2 (complement, slash) is 1, from 1 is 0:

```
          6 7
74 / 5 0 5 4 2
       1̸ 6 1
         2̸ 0
```

Three: 4 × 7 ends in 8, and 8 from 4 is (complement) 6 and slash.

BUT—when we slash a zero, we must always slash the digit to the left of it as well. The 2 to the left of the 0 is already slashed, but we slash it again. We have no choice. If one slash reduces a digit in value by one, two slashes reduce it in value by two, leaving nothing of that double-slashed 2:

```
          6 7
74 / 5 0 5 4 2
       1̸ 6 1
         2̸̸ 0̸ 6
```

Why does it happen this way? The answer is simply that we are multiplying and subtracting at the same time, digit by digit. The use of a complement in either case calls for recording a ten in multiplying (which means canceling a ten here, since we put down only the result of the subtraction) or canceling a ten in subtraction. Now and then, both may affect the same digit—as they did here.

That slashed 0, remember, is now a 9.

Let's go on. See the next answer digit as 8 into 96—MORE THAN TEN.

This signals the need for raising the answer digit by 1. This is not difficult. Just underline it, subtract the divider from the remainder, and the problem now looks like this:

```
          6 7 (7 underlined)
74 / 5 0 5 4 2
       1̸ 6 1
         2̸̸ 0̸ 6
           2 2
```

The remainder 222 comes from the subtraction of 74 from 96—really, of course, 740 from 962.

Now we will find the last digit of the answer. 8 into 22 is 3. Multiply and subtract at the same time on your pad and find out whether or not this problem comes out even.

Do it yourself.

Does the problem come out even or not?

Longer Dividers

Perhaps you are already wondering whether dividing by numbers of three or more digits make things much more complicated.

The answer is, not much.

Nothing in our technique changes one bit, except that we repeat step two of no-carry multiplication as many times as we need to in order to get a full subtraction. Dividing by four- or five-digit dividers is no harder than dividing by two-digit dividers. There are more details and it will take a little longer, but the process is not really different.

You still divide by only the first digit of your divider, raised in value by one, to produce automatically each succeeding digit of your answer. Should that answer digit need revising, that fact will be signaled to you when the remainder is larger than the divider. If you miss that signal, you are notified again when the next answer digit seems to be ten or more.

Once in a very great while an answer digit will need revising twice. After you have raised it once and adjusted your remainder, the remainder will *still* be larger than the divider —and the next answer digit will *still* be ten or more. In such rare cases, just underline the answer digit again (raising it in value by 2) and subtract the divider once again from the remainder before continuing.

The following example is admittedly a wild extreme, so obvious on the face of it that even by rote you could hardly get into this sort of situation. Yet, even should you abandon all your number sense and follow every rule without looking at the problem itself, the rules would eventually bail you out. You would have to raise your answer digit no less than five times, yet it would ultimately be right:

```
           4
           ≡
           =
1 1 / 9 9
      5 5
      4 4
      3 3
      2 2
      1 1
```

This, as we said, is absurd. Yet it demonstrates the absolute reliability of the operating rules even when your own common sense sees nothing wrong. You divide 2 into 9 and see an answer digit of 4, so your first remainder is 55. You raise the answer by one and subtract the divider, giving a remainder of 44. This goes on through four more revisions of the answer digit, the last one taking care of the remainder 11.

The problem was selected especially to demonstrate this ultimate possibility. If an answer digit seems to need revising more than once or twice, sit back and look at the problem as a whole. Chances are you have overlooked something very obvious. The rules are as important to fast mathematics as trees are to a forest—but we take note of the forest first, then use the trees. The folk saying is too obvious to need repetition here.

One more point might be mentioned. If you work entirely by rote, you might sometimes be confused by the placement of your remainder when working out a problem like this:

```
3 2 / 8 9 6
```

The first answer digit, by inspection, is 2—4 into 8. But beware when you start to develop the remainder:

One: 3×2 is in the 0's, and 0 from—

STOP. 0 from 8? This cannot be so. The answer digit would need several revisions. No, in this case it is 0 from 0—the unshown 0 to the left of 8.

Why? Your own number sense should give you a strong inkling. To distill it into an operating rule, it is because you divided into the first *one* digit of the number divided, rather than into the first *two* digits. This in effect moves the product of this first multiplication one place to the left.

So when you divide the first digit of the divider into the first digit only of the number divided, start right out by ignoring what that first answer digit and the first digit of the divider would "be in." They wouldn't "be in" anything but the zeros, or you would have divided into the first *two* digits of the number divided.

The rest of this particular problem, by the way, shows a typical example of two-digit revisions. Let's go through it. So far, the figures look like this:

```
          2
32 / 8 9 6
     2 5
```

The next answer digit, by inspection, is 6—4 into 25. Put it down and work out the remainder:

One: 3 × 6 is in the 10's, and 1 from 2 is 1.

Two: 3 × 6 ends in 8. 2 × 6 is in the 10's. 8 and 1 is 9, and 9 from 5 is—complement, slash:

```
        2 6
32 / 8 9 6
     2 5
     1̸ 6
```

Three: 2 × 6 ends in 2, and 2 from 6 is 4:

```
        2 6
32 / 8 9 6
     2 5
     1̸ 6 4
```

Look at the remainder. Your growing number sense might show right away that it is exactly twice the divider. If not, you would at least notice that it is larger than the divider, so you underline the 6 to raise it to 7, and subtract the divider from the remainder: 3 from 6 is 3. 2 from 4 is 2.

Look at the remainder again. It is the same as the divider. Underline the underlined 6 once again, subtract the divider, and the problem comes out even. The answer is 2 $\underline{\underline{6}}$, which you read or rewrite as 28.

Get out your pad now, have an absolutely clean page on top, and start one problem with a five-digit divider. This one problem embodies just about every possible wrinkle in short-cut division.

$$36,182\,\overline{/\,18,926,824}$$

This problem is bound to be a bit tedious, in any method of arithmetic. We can get a rapid estimate, as the next chapter will show, but for a complete answer there is no avoiding a number of steps.

Take a deep breath and plunge in.

4 into 18—our way of producing the first answer digit automatically—is 4. This 4 goes over the sixth digit of the number divided, since there are five digits in the divider and we started by dividing into two digits rather than one. Now we start on the remainder:

One: 3 × 4 is in the 10's, and 1 from 1 is 0.

Two: 3 × 4 ends in 2. 6 × 4 is in the 20's. 2 plus 2 is 4, and 4 from 8 is 4.

(We are still spelling out every step to make it clear. But as you practice the smooth handling of these steps, your eye and mind should begin to jump from one fact to the other almost without intermediate thought. Step two above will, with experience, become "2—4—4.")

Three: 6 × 4 ends in 4. 1 × 4 is in the O's. 4 from 9 is 5.

Four: 1 × 4 ends in 4. 8 × 4 is in the 30's. 4 and 3 are 7, and 7 from 2 is (complement) 5 and slash left.

Five: 8 × 4 ends in 2. 2 × 4 is in the 0's. 2 from 6 is 4.

Six: 2 × 4 ends in 8. 8 from 8 is 0.

If you followed these steps on your pad, here is what your work should look like:

$$\begin{array}{r} 4 \\ 36,182\,\overline{/\,18,926,824} \\ 4\ \not{5}\ 5\ 4\ \ 0\ \ \end{array}$$

Are you ready to go on? Or would it be a good idea to

take another look at the remainder? The remainder is larger than the divider. So underline the 4—raising it to 5—and subtract 36182 from 44640, left to right, canceling in the answer:

```
                           4̲
3 6, 1 8 2 / 1 8, 9 2 6, 8 2 4
                 4  5̸ 5 4  0
                 1̸  8 4̸ 6̸  8
```

If 1̸84̸6̸8 is smaller than 36182, we are ready to go on. Divide 4 into 8 and put down 2 as the second answer digit. Start developing the remainder:

One: 3 × 2 is in the 0's. Ignore it, because we divided into one digit rather than two.

Two: 3 × 2 ends in 6. 6 × 2 is in the 10's. 6 plus 1 is 7, and 7 from 8 is 1.

Three: 6 × 2 ends in 2. 1 × 2 is in the 0's. 2 from 3 is 1.

Four: 1 × 2 ends in 2. 8 × 2 is in the 10's. 2 plus 1 is 3, and 3 from 5 is 2.

Five: 8 × 2 ends in 6. 2 × 2 is in the 0's. 6 from 8 is 2.

Six: 2 × 2 ends in 4. 4 from 2 (complement) 8, and slash.

Check your jottings against the model at this point:

```
                           4̲ 2
3 6, 1 8 2 / 1 8, 9 2 6, 8 2 4
                 4  5̸ 5 4  0
                 1̸  8 4̸ 6̸  8
                    1 1 2  2̸ 8
```

This should be enough step-by-step explanation. Go ahead and finish this problem. You will find that it does not come out even. There will be a remainder. Make sure to examine the remainder carefully. There is a reason why you should.

Do not look ahead to the finished problem until you have a remainder that satisfies you. Then, if you feel you understand all the techniques in this chapter, go ahead to build speed in division.

The remainder, as you undoubtedly discovered, held within it a revision of the last digit:

```
                        4 2 2
                        _   _
36,182 / 1 8 , 9 2 6 , 8 2 4
           4   5 5 4   0
           1   8 4 6   8
               1 1 2   2 8
               1 4 0   8 2 0
                   3   7 4 8
```

Final answer, 523; remainder, 3,638.

9

BUILDING SPEED IN DIVISION

THE last exercise in the last chapter was a stunner. It was, just from the quantity of digits to be handled, the most tedious situation you are likely to face in arithmetic. The numbers go on and on, and if you need a full answer there is simply no way to avoid dealing with every single digit.

Most division is much simpler.

First of all, we seldom need to work out any division problem of this length in such detail. Largely because American business has become accustomed (and wisely so) to dealing in rounded-off numbers, you would most likely find such a problem rounded off to start with. Second, the only reason we had to use every single digit was to get a fully accurate remainder. We would have got precisely the same whole-number answer by cutting down the divider from five to three digits.

Remember the two reasons for working from left to right: it is more natural, and it is also self-estimating. Turn back to the last problem for a moment, then compare the final working with this version of *the same problem*:

```
              4 2 2
         ___________
3 6 2  / 1 8 9 3 6 8
           4 5 5
           1 8 3
             1 1 2
             1 4 0 4
               1 4 2
```

This is really the same problem—but a rounded-off version of it. We rounded it off before beginning by doing two things:

First, we rounded off the divider to three digits because we saw simply by inspection that the whole-number answer would be in three digits. In rounding off, the first three digits (361) became 362 because the following digit is 5 or more.

Second, we dropped the *same* number of digits from the number divided as we did from the divider. This ensures that our answer will not be ten times too big or ten times too small.

Rounding off such a problem is obviously faster as well as simpler. For most purposes, it is quite accurate enough. You will note that the remainder is not the same, and sometimes the last digit itself might be off by one or two points in value—but we are still more accurate than a slide rule.

When you need a very quick estimate, you can carry this even further. If you care only about the first two digits of the answer, then round off your divider to two digits and cross out as many digits in the number divided as you did in the divider.

In this event, the full divider 36182 becomes 36. The number divided, 18936824, becomes the far more manageable 18937.

Your solution now looks like this:

```
          4 2 5
      ___________
3 6 / 1 8 9 3 7
        4 5
        1 9
          2 1
          1 3 9
```

Notice that the third digit of your answer is no longer accurate at all. But your first two digits are.

You would seldom simplify a problem to quite this extent, because the possible error is ten to twenty per cent, but it is a useful device to know when speed rather than perfect accuracy is required for a very fast approximation.

Try rounding off one sample to make sure you have the idea firmly in mind—especially the proper handling of the number divided. Reduce this example to a form suitable for a three-digit answer:

$$68,392 \,/\overline{268,595,471}$$

As we pointed out before, an answer of which the first two digits are correct, and the third digit is one more or one less than it should be, can never be more than one per cent wrong, and may be as little as one-tenth of one per cent wrong. The least error is 998 when it should be 999.

Your rounding off of the above problem should look like this:

$$684 \,/\overline{2,685,955}$$

In this case, both final digits were raised by one in the rounding off, because both following digits were 5 or more. The only new element in this kind of rounding off is establishing the proper size of the number divided. Since we dropped two digits from the divider in this case, we also dropped two digits in the number divided.

Other than some of the short cuts in the last part of this book, which apply to many (though not all) problems, this is about all there is to know about estimating in division. The most important elements, as you can see from the work so far, are your quickness and confidence with the basic digit combinations in dividing, multiplying, and subtracting, and your mastery of the one-two-three of shorthand division.

It is now time to brush up on your vocabulary.

Division is, after all, only multiplication done backwards. Instead of "seeing" 6×7 as 42, we learn to see $6 \,/\overline{42}$ as 7 . . . or $7 \,/\overline{42}$ as 6. Just as it is in adding, subtracting, and multiplying, the best medicine for this is repetition.

Keep in mind that the following practice section is *not* to be done as a simple division drill. Go slowly and carefully, making every effort to "see" the answer rather than the problem. It may help to say aloud the answer in each case, shoving the problem as far back in your mind as you can.

Once again, you are practicing to see *h* and *e* as "he"—not as "*h* and *e* spell 'he.' " You can do this with numbers just as you can with letters if you spend a reasonable amount of time at it.

Try to read through these just as if they were words, seeing the words rather than the letters:

$3\overline{)3}$	$8\overline{)32}$	$8\overline{)48}$	$7\overline{)14}$
$6\overline{)42}$	$9\overline{)72}$	$3\overline{)21}$	$5\overline{)20}$
$9\overline{)81}$	$2\overline{)16}$	$9\overline{)9}$	$1\overline{)4}$
$6\overline{)6}$	$3\overline{)9}$	$4\overline{)40}$	$4\overline{)8}$
$7\overline{)49}$	$1\overline{)5}$	$9\overline{)27}$	$1\overline{)9}$
$8\overline{)72}$	$6\overline{)30}$	$4\overline{)24}$	$2\overline{)4}$
$2\overline{)6}$	$1\overline{)2}$	$2\overline{)10}$	$7\overline{)63}$
$7\overline{)35}$	$8\overline{)16}$	$6\overline{)18}$	$5\overline{)10}$
$3\overline{)24}$	$4\overline{)4}$	$4\overline{)32}$	$1\overline{)8}$
$4\overline{)16}$	$1\overline{)6}$	$5\overline{)45}$	$9\overline{)45}$

This is quite a new bit of practice for most of us. Even though division is merely inverted multiplying, it is the basic process on which the average person has spent less "drill" time learning his tables than on any other. Yet, for quick working of short division (or even long division), there is no substitute for knowing them backwards and forwards.

Your confidence and accuracy with any method of speed mathematics are based entirely on your confidence and accuracy with the individual digit combinations. No technique can be very helpful in your daily mathematical needs unless you can *do* it—with confidence and accuracy.

Improve your handling of division now by practicing the

rest of the possible combinations. As always, work at seeing only the answer—not the problem:

8 / 24	9 / 18	8 / 56	5 / 30
2 / 8	3 / 15	4 / 12	2 / 18
9 / 63	8 / 8	2 / 14	7 / 21
1 / 3	6 / 36	8 / 40	9 / 36
5 / 35	4 / 28	5 / 5	4 / 20
7 / 42	6 / 18	7 / 28	6 / 54
4 / 36	8 / 64	3 / 6	3 / 12
5 / 25	5 / 15	6 / 48	9 / 54
3 / 18	7 / 7	2 / 2	2 / 12
6 / 24	7 / 56	1 / 7	3 / 27

That is the whole series. There are no other combinations.

There is, though, an important variation. When you stop to think about it, division is the only one of the four processes in which you usually have an *approximate* answer.

When you add, you get one specific result. 9 plus 6 is always (whether you add it or whether you subtract a complement and record a ten) 15.

When you subtract, there is no question about it. 8 from 13 is always 5.

When you multiply, 4 times 7 is always 28. There are no if's, and's, or but's about it.

But what about 8 / 3 1 ?

Your instinct or number sense or practice at division tells you that the answer to this is "almost 4." But it is not 4. No matter how close it is, you still cannot get four 8's into 31.

It is so close, of course, that you can get 3 and 87/100's 8's into 31. But you still do not get 4.

You will get 3 +. Your answer will approach 4 as you work out the remainder in decimal or fractional form, but *your first digit has to be 3*. This is because our methods of writing numbers include ways to write 3 plus a fraction, but not 4 minus a fraction.

The thought is worth considering because quick and efficient division requires us to "see" 8 into 31 as 3.

The usual process for many of us is to take a stab at the closest answer, then (consciously or unconsciously) multiply it out in our minds to see if it checks out, and revise our trial digit when required.

The automatic digit-finding technique of shorthand division (dividing by only the first digit of the divider, increased in value by one) solves a large part of the problem. The second half of the battle, however, is to learn to "see" an approximate division, such as the one above, cleanly and properly at first glance. This means knowing at sight that $8\ /\overline{47}$ is "5," even though the final answer will be much closer to 6.

Here is some practice on this, which will pay in faster and easier dividing. Work at these tables with the objective of "seeing" only the *first* answer digit. Do not worry about whether the eventual answer will be 3.001 or 3.999. In both cases, you start with 3.

Start now:

$6\ /\overline{59}$	$8\ /\overline{15}$	$6\ /\overline{29}$	$3\ /\overline{8}$	$2\ /\overline{15}$
$9\ /\overline{71}$	$8\ /\overline{31}$	$4\ /\overline{15}$	$3\ /\overline{23}$	$7\ /\overline{34}$
$2\ /\overline{5}$	$8\ /\overline{71}$	$7\ /\overline{48}$	$6\ /\overline{5}$	$9\ /\overline{80}$
$6\ /\overline{41}$	$3\ /\overline{2}$	$9\ /\overline{44}$	$5\ /\overline{9}$	$7\ /\overline{62}$
$2\ /\overline{3}$	$4\ /\overline{7}$	$5\ /\overline{19}$	$7\ /\overline{13}$	$5\ /\overline{44}$
$4\ /\overline{31}$	$6\ /\overline{17}$	$2\ /\overline{9}$	$4\ /\overline{23}$	$9\ /\overline{26}$
$4\ /\overline{39}$	$3\ /\overline{20}$	$8\ /\overline{47}$	$7\ /\overline{55}$	$5\ /\overline{14}$
$8\ /\overline{63}$	$8\ /\overline{79}$	$4\ /\overline{27}$	$6\ /\overline{35}$	$3\ /\overline{14}$

This group includes over half the possibilities. Perhaps you have seen the nature of the practice you are now doing. Each of the division examples you did contains a number divided just 1 less in quantity than one which would call for a higher first-answer digit. For instance, $3\ /\overline{14}$ is practically 5—but you start with 4.

Learning to "see" what we might call the breaking point

of each answer digit cannot help but ease and speed up your automatic division. Once you have learned to "see" $3\overline{)14}$ as 4 rather than "almost 5," you should have no trouble reading $3\overline{)13}$ as 4 also. If you recognize $8\overline{)71}$ as 8 rather than "almost 9," then whenever you need an answer digit for $8\overline{)70}$, $8\overline{)69}$, and all the intermediate possibilities down to $8\overline{)64}$, you should answer without a second thought "8."

Go through the rest of these "breaking point" combinations now:

$9\overline{)17}$ $6\overline{)23}$ $3\overline{)17}$ $5\overline{)24}$ $4\overline{)35}$

$7\overline{)41}$ $5\overline{)34}$ $9\overline{)62}$ $2\overline{)7}$ $8\overline{)23}$

$3\overline{)26}$ $2\overline{)11}$ $9\overline{)53}$ $3\overline{)11}$ $6\overline{)53}$

$4\overline{)19}$ $9\overline{)35}$ $7\overline{)20}$ $2\overline{)17}$ $9\overline{)89}$

$6\overline{)47}$ $3\overline{)5}$ $7\overline{)69}$ $5\overline{)49}$ $9\overline{)35}$

$3\overline{)29}$ $2\overline{)19}$ $5\overline{)29}$ $6\overline{)47}$ $7\overline{)27}$

$8\overline{)39}$ $2\overline{)13}$ $4\overline{)11}$ $8\overline{)55}$ $5\overline{)39}$

That finishes all the possibilities. This practice section is vastly different from any other in the book, in that it calls for you to give what you know to be a very *approximate* answer—an answer you know full well will need adding to later. Yet this is the way we must divide, and what at first may seem very odd must become second nature.

Having done the single-digit tables, expand your practice a bit now by doing precisely the same thing with these examples. In each case, be sure to divide by only the first digit of the divider, raised in value by one. See only the first answer digit for each of these problems:

$67\overline{)498}$ $24\overline{)918}$

$862\overline{)368215}$ $46\overline{)111}$

$93\overline{)4863}$ $3486\overline{)64162}$

$74\overline{)3649}$ $62\overline{)3524}$

$555\overline{)521893}$

Those are the basic vocabulary elements in speed division. When you put them together with simultaneous left-to-right multiplication and subtraction, short hand division really becomes short-cut division.

See how well you remember the entire system now by working this problem out in detail on your pad:

```
652 / 24776
```

Cover the demonstration below with your pad while you work it out. This problem comes out even, so you know without looking ahead whether or not you solved it properly.

When you are finished, compare your working with this model:

```
              3 7
               ‾
652 / 2 4 7 7 6
      1̸ 5 2 1
        1̸ 7̸ 5 2
```

One way to speed up your working of any problem such as this is simply to jot down your working figures on whatever piece of paper you have handy. You do not always have to copy the entire problem, although this is often helpful during early stages of practice, when each move still seems rather strange.

If you have been doing your practice, you should be able to solve the following problem by the shorthand method without copying it. I do not suggest that you solve it in your head (some people can, but most of us have to lean on our pencils even with simplified techniques), but you should be able to glance from printed problem to jotted working figures and produce the answer without turning yourself into a stenographer:

```
46 / 40158
```

See if you can do this problem without copying it. Use your pad only for jotting down the answer and working figures. If you have never done it this way before the technique may seem difficult, but it can save you a great deal of time in your number work.

When you have finished, compare your jotted figures with these:

Answer—8 6 2

Working
figures – ~~1~~ ~~4~~ 3
~~1~~ ~~6~~ 9
1 3
~~1~~ 4 6

Try a simpler problem with this non-copying method. See if you can jot down your answer and your working figures for this example without coyping the problem itself:

$7\,\overline{)\,2\,4\,3\,6}$

Cover the answer below until you have done your best.

Here is the way you jot down your answer and working figures:

Answer—3 4 8

Working
figures – 3
~~1~~ 5

You have practiced all possible digit combinations. Now, before going on to other methods of speeding up your number work, do a bit more drill in actual division examples.

If you can, solve these without copying them. Use your pad for answers and working figures only.

$8\,\overline{)\,5\,0\,8\,0}$ $6\,\overline{)\,2\,6\,1\,6}$ $9\,\overline{)\,8\,8\,8\,3}$

$46\,\overline{)\,3\,1\,2\,8}$ $37\,\overline{)\,9\,2\,5}$ $69\,\overline{)\,3\,7\,2\,6}$

$58\,\overline{)\,3\,9\,6\,7\,2}$ $34\,\overline{)\,1\,4\,6\,8\,8}$

$75\,\overline{)\,6\,7\,3\,5\,0}$

If you did the above examples in the shorthand method with confidence, then divisions of any length at all are merely extensions of what you already know. Try the following three-digit dividers. Again, do your best to jot down only the developing answer and working figures rather than copying the problem over:

```
674 / 218376      468 / 373464

198 / 125136      837 / 185814
```

Give these your best before checking your work against the solutions following. If you were able to handle them without copying, extra good. The solutions will be given in copied form, however, so you can check your work whether or not you jotted your answer and working figures separately.

Every one of the preceding problems comes out even, as a quick check on yourself before looking at the solutions. If you have any remainders, go back and recheck now.

```
      6 3 5           4 3 6           9 8 7
8 / 5 0 8 0     6 / 2 6 1 6     9 / 8 8 8 3
      2               2               7
        4             1 3               6

        6 7             2 5            5 3
46 / 3 1 2 8    37 / 9 2 5     69 / 3 7 2 6
     1 4 6           2 8               3 7
       1 4 6                           1 7 9

            6 8 3                      3 2 1
    58 / 3 9 6 7 2          34 / 1 4 6 8 8
             5 8                 1 4 4
               2 3                 1 0
             1 6 8                 1 4 0
                                     1 6
                                       3 4

                     8 9 7
            75 / 6 7 3 5 0
                 1 7 3
                   1 6 0
                     2 8 5
```

```
                 3 2 3                              7 9 7
            _____________                      _____________
6 7 4 /  2 1 8 3 7 6           4 6 8 /  3 7 3 4 6 4
            1 1 6 1                              1 5 6 8
                3 7 9                                1 4 7 4
                1 6 7 4                                  1 5 7 8

                 6 3 2                              2 2 1
            _____________                      _____________
1 9 8 /  1 2 5 1 3 6           8 3 7 /  1 8 5 8 1 4
            1 1 7 3                                  2 8 4
                1 4 9                                  2 7 7
                                                       1 8 4 7
```

10

ACCURACY: THE QUICK CHECK

PRODUCING a quick answer is not always the end of a problem in arithmetic. A wrong answer can be worse than none at all.

When you balance your checkbook, you care whether every stub was done perfectly because otherwise you face an unpleasant hour or so finding out why your balance does not agree with the bank statement. When you make out your income-tax return, you check and double-check every operation to make sure you are not paying too much, or else getting yourself into trouble with wrong arithmetic. And in business, where so many decisions are based on numbers, the wrong numbers can lead to wrong decisions.

There are two fast ways of checking your answers. The faster of the two is the subject of this chapter. A slightly more complex "back-up" check is discussed in the next chapter. Both of them are infinitely quicker than the standard technique of doing the problem over.

The standard way of checking an answer is effective, but very slow. It takes just as much time to tell whether an answer is right as it does to produce it in the first place. The standard way, of course, is to do the problem over again in the opposite way. If we got the answer by adding down, we check by adding up. This (to some extent) keeps us from repeating some habitual error that we might commit twice if we handled the figures in the same order both times. If we subtracted, we check by adding the answer and the smaller number to see if the total equals the larger number. If we multiplied 897 by 123, we check by multiplying 123 by 897. And if we divided, we check by multiplying the answer by the divider, adding the remainder, and seeing if the result equals the number divided.

There are two serious weaknesses to this "backward" method of checking.

First, it is slow and rather boring. Our object is speed and accuracy, with as little boredom as possible.

Second, it is not really a proof at all. If we get the same answer both times, we *assume* the first solution to be correct. Yet if we habitually think of 4×7 as 32 (and such habitual mistakes are not uncommon), then we might indeed get the same wrong answer twice. Even if our second try produces a different answer, we still do not know if one of them is right —or which one it is. We must do the problem still a third time.

The techniques of simplified mathematics you have learned are inherently more accurate (because they are simpler) than traditional methods, but it is still unwise to assume an answer is correct unless you *know* it is correct because you have checked it.

The two methods for checking answers you are about to learn are very similar. Neither one is new, although some of the short cuts in applying them are. The first method is known in mathematical circles as "casting out nines" or "the digit sum" method. The second is "casting out elevens." Both work on a system of check figures completely divorced from your calculations in solving the problem, so habitual errors are unlikely to be repeated, but the methods of deriving the check figures are quite different. Actually, each of them is a way of testing whether the remainders of nine or eleven remain properly constant through your calculations. This will be discussed at greater length in the next chapter. First, learn the technique of handling what is known as the digit sum.

The Digit Sum

The digit sum, as the phrase suggests, is simply the sum

of all the digits in a number. This sum will be your "check figure" for each number.

Learn first how to find a digit sum. Then we will go on to the ways of using it. After you have found a few digit sums, you will be able to derive one almost as fast as you can read the number itself. It is really that quick.

If the digit sum is merely the sum of the digits in a number, then the digit sum of 23 should be 2 plus 3, or 5. Odd as this may seem at first, that is precisely right. The digit sum of 23 is 5.

The digit sum of 341 is 3 plus 4 plus 1. The digit sum of 341, then, is 8.

Just add the digits. The digit sum of 42 is—

Did you get 6?

Now, however, it becomes a little trickier. For quick utility, the check figure must always be a single digit. But the sum of the digits in longer numbers goes over ten.

In this case, we use the digit sum of the digit sum. This is the digit sum of the number itself.

This is how it works. The digit sum of 587, for instance, goes into two digits by the time we add 5 and 8, which make 13. When we add the final 7, we have a digit sum of 20.

You can reduce this to a single digit at the end, by adding 2 plus 0 and getting 2. Or you can *reduce as you go along,* like this: 5 plus 8 is 13. Reduce this by adding 1 plus 3 to get 4. 4 plus the final 7 is 11. Reduce this by adding 1 plus 1 and get 2.

This peculiarity of the digit sum is only a foretaste of those to come. Let us finish this thought before getting to that, however. Try one digit sum now. Add all the digits of the number 6934 and then add the digits of the answer until you come out with a single digit. Then reduce as you go along through the same number 6934 and see if you come out with the same final digit sum.

Done the first way, you add 6 plus 9 plus 3 plus 4 and get 22. Reduce this by adding 2 plus 2 to get 4. The digit sum of 6934 is 4.

Done the second way, you add 6 plus 9 to get 15. The digits of this total 6. 6 plus 3 is 9, plus 4 is 13. 1 plus 3 is 4. The digit sum is still 4.

There is, however, a third way. This third way is called casting out nines. The reason for the name is inherent in the digit sum, and is a fascinating byway in the mysteries of numbers.

The odd fact boils down to this: If you divide any number by nine, the remainder is the same as the sum of all the digits of that same number—reduced to one digit by continually adding the digits of the sum of the digits until you wind up with one digit.

In other words, the digit sum of any number divisible by nine will be nine. The algebraic proof of this is a little complicated for this book, but you can demonstrate it for yourself.

Take one of our examples of a minute ago. We found that the digit sum of 587 is 2. If you divide 587 by 9, you will get an answer of 65—and a remainder of 2.

The last example we tried was 6934. Our digit sum was 4. Try dividing 6934 by 9. The answer is 770—and a remainder of 4.

The digit sum is, in essence, the same as the "nines remainder"—the amount left over after an even division by nine. This is important not only to digit sums but in understanding how the entire check-figure system works, A more complete explanation comes in the next chapter.

The fact that the digit sum is the same as the remainder after dividing by nine brings up two more useful oddities. First, nine (for digit-sum purposes) becomes zero. Second, a digit 9 counts for nothing in the number itself.

This brings up a great short cut in deriving digit sums. As you add the digits, simply ignore any nines. They do not count.

Demonstrate this to yourself a few times. The thought takes a little getting used to.

Add the digits of 19. The total is 10. The digit sum of of this is 1. If you looked at 19 and ignored the 9, you would see 1 anyway.

Now try 29. 2 plus 9 is 11, which reduces (1 plus 1) to 2. Look at the same number, ignoring the 9, and you see 2.

See if you can find *any* combination of two digits, of

which one is 9, which, when you add the digits and reduce, does not produce the digit which was not 9. This is an intriguing and frustrating search. 95 becomes 14, which reduces to 5. 89 becomes 17, which reduces to 8. 93 becomes 12, which reduces to 3.

Do not stop with two-digit numbers. Try *any* number you wish, that contains a quantity of 9's and any other digit. Convince yourself of this very peculiar truth by reducing these numbers to digit sums:

1999	9949	969
399	9299	99989

This is a strange phenomenon, but in addition to being strange it is highly useful. It means that in finding a digit sum your eye can simply skip over any 9's. They will not change the digit sum. The digit sum of 99999999999997 will be 7.

Perhaps your mind is already ranging ahead, wondering if digits in a number that add up to 9 behave in the same way. If the digit 9 does not change the digit sum, what about 3 and 6?

Try it and see. Find the digit sum of 361. Actually work it out. Now envision the 3 and the 6 as adding to 9, and therefore to be ignored. Cast both of them out, as you would cast out a 9. Your answer, of course, is 1—the 1 you see if you ignore the 3 and the 6 (because they add to 9) in 361.

The lesson is quite true. Since 9 will not affect the digit sum, you may ignore any 9's you see in the number—or any combination of digits that add to 9.

Try these:

145	727	463	273
254	381	574	186

In each case, you will find that adding all the digits and then reducing by adding together the digits of the sum (as many times as you need to) is precisely the same as the digit left after casting out digit combinations that would add to 9 —no matter where those digits appear in the number.

Zeros, too, obviously count for nothing. You would not

add them as you added the digits anyway, so you can safely ignore any 0's in any number as you derive its digit sum.

For digit-sum purposes, 9 and 0 are equal. This is only a device for this particular purpose, of course. But for simplicity in working, consider a final digit sum of 9 to be 0. It would come out to the same result in the end, and it can save a significant amount of time to wipe out the 9 to start with.

Before you learn how to apply digit sums in checking your results to problems, try deriving a few. Ignore any 0's, 9's, or combinations of digits adding to 9 in the following numbers as you extract the digit sum of each:

16428	32,718,643
73619	84,600,372
24583	26,738,514

Notice that one of the digit sums above works out to 9. This, for digit-sum purposes only, can be treated as 0.

Running Adjustment

One more short cut is worth noting in developing digit sums. Since you know that 9 or any combination of digits adding to 9 (such as 324) can be ignored, you can also think of any pair of digits adding to ten as being worth 1, or any pair adding to 8 as subtracting 1 from the partial total already in your mind, and so on.

Glance back at the first example above. You can do it with extra speed by counting "1—(6 and 4 are complements, count as 1) 2—(2 and 8 are complements, count as 1) 3."

In the last example, you might start adding like this: "(2 and 6 are 8, or minus 1) from 7 is 6—and 3 is 9, or 0—(8 is minus 1) from 5 is 4—and 5 (group 1 and 4) is 9, or 0. Digit sum, 0."

I think you already see how you will soon be able to derive the digit sum of a number almost as fast as you can read the number itself. You simply add up to 9 and then start over, dropping each 9 in turn and not even recording it. In doing so, you use every trick of grouping you have learned.

These extra-speed tricks are helpful to very rapid work. They can become so fast and so easy you could, if you wish, make a parlor trick out of the idea. Glance at any figure and predict the total of its digits, totaled in turn until you get a single digit. You will have your result, if you play with these methods a bit, before your challenger has added the first three digits.

Try it once on this number:

869,325,008,462,118

Watch how quickly it goes: "8 from 6 is 5—skip 9—plus 5 (the 3 and 2) is 1 (ten reduced)—plus 5 is 6—skip the 0's—minus 1 is 5—plus 1 (the 4 and 6) is 6—plus 4 (the 2, 1, 1) is 1 (ten reduced)—less 1 (the final 8) is 0."

After a few more moments of practice, you will find yourself almost scanning a digit sum. You will ignore pairs adding to 9. You will add 1 for pairs that are complements, and subtract 1 for 8's or pairs adding to 8. Beyond this, you may begin to note pairs adding to 7 (or 7's themselves) as subtracting 2. You may even begin to skip around a little, "seeing" 485 in a long number as minus 1 because the 4 and 5 add to 9 and the 8 is minus 1.

This is such a joyous and useful byway of numbers that you will profit by making a game of finding digit sums as quickly as you can.

Checking Your Answers

The digit sum is not merely fascinating. Its utility is in the quick check.

The general rule for checking by digit sums is simply this: Do to the digit sums of the numbers in the problem whatever you did to the numbers themselves. The result must equal the digit sum of the answer—if the answer is correct.

If you add a column of numbers, then you simply add the digit sums of those same numbers. This result (reduced as always to a single digit) must equal the digit sum of the answer. If you multiply two numbers, then you multiply their digit sums. This, reduced, must equal the digit sum of the correct answer.

The reason why it works will be explored in the next chapter. For the moment, let us see how it works.

Follow this example in addition:

	Problem	Check
	146	2
	928	1
	357	6
Totals	1431	0

Digit sum of answer: 0

In this case, the sum of the digit sums is the same as the digit sum of the answer. Check.

Once you are in full training at digit-sum reduction, you will be able to check such a problem about as fast as you read it over. A peculiarity of checking problems in addition, especially, is that since you added the numbers you can merely add all the digit sums in one operation. That is, you can develop one digit sum for the columns of digits in one operation instead of getting a separate sum for each number. In the problem above, it would be equal to getting a digit sum for 146,928,357. If you try it, you will find that this digit sum is 0.

Now for a longer problem. Each digit sum appears on a separate line for clarity, but you do not need to do it this way. You can go through the three numbers one after the other until you have one final digit sum—which in this case will be 3:

	Problem	Check
	68,352	6
	97,834	4
	35,876	2
Totals	202,062	3

Digit sum of answer: 3

Now try these problems and check your answers by using

digit sums. Be sure to work at your new habits: work from left to right, use complements, and record tens:

18	73	4,832
26	82	4,689
54	65	2,234
93	20	5,367
21	37	

Cover the answers and their check figures with your pad until you have finished.

Here are the totals, together with their digit sums:

212 (5) 277 (7) 17,122(4)

Locating Errors

Before going on to the ways of using digit sums in other types of problems, face one imperfection in the system—and learn a special advantage in return.

Digit sums do not invariably catch every type of error. The errors they miss are so unlikely that for all practical purposes you can almost forget them, but you should know about the possibility.

Since for digit-sum purposes 9 is the same as 0, you can easily see that this method of checking will not catch an error in which one digit in your answer is 9 when it should be 0, or 0 when it should be 9. If you have two correct digits, but have them reversed (36 instead of 63) it will not catch this either. Or if by any odd chance your error consisted of a digit or combination of digits that was exactly 9 more or less than it should be, the digit-sum check would not ferret this out either.

Actually, years of experience have shown that the errors not caught by the digit sum are exceedingly rare. For most needs, it is perfectly adequate—far more accurate than doing the problem over, in fact.

In return for these shortcomings, however, the digit-sum check offers a substantial bonus.

The digit sum will not only tell you if your answer is

wrong; it will also tell you *how much it is wrong*. If the digit sum of your answer is 4, and you find that it should be 7, then you know that one digit of your answer is too low by exactly 3. You do not know which digit it is, but the fact that one digit is precisely 3 less than it should be is helpful in locating the error quickly.

Checking Subtraction

Our general rule is that you do to the digit sums of the numbers whatever you did to the numbers themselves. This result, reduced, must equal the digit sum of the correct answer.

In subtraction, it is important to recall that for digit-sum purposes we can consider 9 to be 0. This is because you will sometimes have to subtract a larger digit sum from a smaller. The way to do it is to add 9 to the digit sum that is otherwise too small to be subtracted from.

Here is an example of this situation:

Problem	Check
615	3
− 593	8
22	8 from 3 won't work. But 8 from 12 (3 plus 9) is 4.

Digit sum of answer: 4

You do not always have to add 9 to one digit sum before you can subtract the other. About half the time, the digit sum of the larger number will be as large as or larger than the digit sum of the smaller number. In this case, of course, you do not tamper with either digit sum; you simply subtract.

Another way to tackle the check when the situation is as above is not to subtract at all. You will get exactly the same result by *adding* the digit sum of the answer to the digit sum of the smaller number. This, if the answer is correct, must equal the digit sum of the larger number. Try it on the example above: The digit sum of the answer (4) plus the digit sum of the smaller number (8) is 12, which reduces to 3. This is the digit sum of the larger number. Check.

Try these subtractions and check them with digit sums. Remember to work from left to right, use complements, and cancel in the answer:

7,382	1,123	586,493
− 6,987	− 1,099	− 465,906

Because finding digit sums themselves is and should be entirely a mental process, you may not have used your pad recently. Locate it now and actually do the above problems and their digit-sum checks before uncovering the answers below.

Now compare your results with these:

395 (8) 24 (6) 129,587 (5)

If you have any difficulty in determining how the digit sums of the numbers in each problem worked to produce the digit sum of each answer, go back over the last two or three pages. You cannot subtract 3 from 2—but you can subtract 3 from 11, or add 3 and 8 to get 11, which reduces to 2.

Checking Multiplication

In checking multiplication, you follow the same general rule that applies to all digit-sum proving: since you multiplied two numbers, you multiply their digit sums. This result, reduced to a single digit, must equal the digit sum of the correct answer.

Here is an example:

Problem	Check
421	7
× 17	8
7,157	7 × 8 is 56. This reduces to 11, which reduces to 2.

Digit sum of answer: 2

Odd as it may seem to multiply digit sums together, that is just what you do in order to prove multiplication. As you can see, it works.

Suppose, though, that you set out to check a multiplication and found this result:

Problem	Check
568	1
× 4	4
2,372	4

Digit sum of answer: 5

Something is wrong. The product of the digit sums does *not* equal the digit sum of the product.

The key here is that the digit sum of the answer is 1 higher than it should be—if the digit sums of the individual answers are correct. If the digit sum of the answer is 1 higher than it should be, then one digit of the answer is 1 higher than it should be, too.

Does this help you locate the error more quickly than you otherwise would? Try it and see. One digit of the answer is exactly 1 higher than it should be.

Try the two following examples on your pad, covering the answers below with the pad until you are finished. Work from left to right with the no-carry method, and check your answers with digit sums:

```
  362          874
×  43        × 736
```

In order to check your answers to these problems, of course, you multiply the digit sums and reduce. Here are the results:

15,556 (5)   643,264 (7)

Don't forget that when any digit is multiplied by 0, the result is 0. So if the digit sum of *either* of the multiplied numbers is 0 (or 9) the digit sum of the answer must be 0. For instance:

Problem	Check
3 8	2
× 9	0
3 4 2	0

Digit sum of answer: 0

In a case like this, keep in mind that despite the apparent extra dangers of multiplying by 0 (which would seem to permit any digit sum at all for the other number without changing the final check figure), the answer to the problem must also have a digit sum of 0 in order to check out. So it is as accurate as any other digit-sum proof.

Checking Division

When we come to checking division with digit sums, we have to use a special application of the general rule. Instead of trying to divide the digit sum of the divider into the digit sum of the number divided, work the process in reverse. Multiply the digit sum of the divider by the digit sum of the answer. This, reduced, should equal the digit sum of the number divided.

The reason for this special handling is illustrated by the following example:

Problem	Check	
4	Divider	8
17 / 68	Answer	4
	Number divided	5

Check: 4 × 8 is 32, which reduces to 5. This is the digit sum of the number divided.

You can easily see the trouble you would have trying to divide the digit sum of the divider (8) into the digit sum of the number divided (5) and produce any rational whole-digit result. The reason for this lies in the special reduction of digit sums, which pretends (for digit-sum purposes) that

10 is 1, that 14 is 5, and that 9 is 0. The system works perfectly if you multiply as outlined above, but cannot possibly work if you try to divide.

If there is a remainder in the answer to the division, add one more step. First multiply the digit sums of the divider and the answer, as before. Now, however, *add* the digit sum of the remainder. This total, reduced, should equal the digit sum of the number divided.

Here is how it works:

Problem	Check	
$\begin{array}{r} 2 \\ 23 \; / \overline{48} \\ 2 \end{array}$	Divider	5
	Answer	2
	Remainder	2
	Number divided	3

Check: 2 × 5 is 10, which reduces to 1. 1 plus 2 is 3, which is the digit sum of the number divided. Right.

Now try these problems, using your pad to cover the correct solutions as you always do. Caution: Any digit multiplied by 0 must give 0.

$8\ 2 \; / \overline{6\ 5\ 6}$ $\qquad$ $6\ 2 \; / \overline{8\ 3\ 9\ 1}$

The illustrations below will, as always, be in shorthand division. Look at them after you have finished your practice.

Problem	Check	
$\begin{array}{r} 8 \\ 8\ 2 \; / \overline{6\ 5\ 8} \end{array}$	Divider	1
	Answer	8
	Number divided	8

Check: 1×8 is 8. Right.

Problem

```
          1 3 4
      ----------
6 2 /  8 3 9 1
       2 1
       1 3 3
         1 0 3
           2 1
```

Check	
Divider	8
Answer	0
Remainder	3
Number divided	3

Check: 0×8 is 0. 0 plus 3 is 3. Right.

This is all there is to know about digit-sum checking. The back-up check in the next chapter works the same way, but the check figures will be quite different.

11
ACCURACY: THE BACK-UP CHECK

THE digit-sum, or "casting out nines," method is the quickest and easiest way to check any problem. Once you become fully accustomed to it, you will find yourself checking a problem about as quickly as you could read it over.

It is not, however, completely foolproof. The last chapter explained the types of errors to which it is quite blind. As someone once pointed out, the digit-sum method will tell you that a problem is *wrong,* but it will not tell you for sure that it is right.

This chapter explains how to "cast out elevens." This is a little slower but inherently more accurate than casting out nines. In cases of critical accuracy, some experts advise using both methods. You can easily do one right after the other in much less time than it would take to check by conventional methods, and if both your digit-sum and your "elevens" results check out, you can be quite sure you have a perfect answer.

Casting out elevens, or simply "elevens" as we will call it, works on precisely the same check-figure method as does casting out nines. In fact, adding up the digits is really only casting out nines because the proof of a number's divisibility by nine is the addition of its digits. If the sum is nine (or 0), the number is exactly divisible by nine. Any other result is the *remainder* you will have after dividing by nine.

Both casting out nines and casting out elevens are merely special (and convenient) applications of a general rule. You could check a problem by "casting out" any number at all. You could find the remainder of each number after dividing it by four, say, and use these remainders as check figures. Nines and elevens are merely the easiest numbers to cast out that also depend for their divisibility on every digit in the number.

This use of a division-remainder is not as odd as it might sound at first. If you add a series of numbers exactly divisible by four, then their total must obviously be divisible by four. If one of those numbers has a remainder of two after a division by four, then the answer must also have a remainder of two after a division by four. If you multiply two numbers each of which is exactly divisible by seven, then their product must also be exactly divisible by seven.

When the numbers are not exactly divisible by whatever number you use for your check figure, then the remainders of each number get carried along through the arithmetic too, and once you do to these remainders whatever you did to the numbers themselves, they must come out in exactly the same relationship to the remainder of the answer.

In order to get a clearer understanding of what is behind this general method of checking, try "casting out" the fives in the following example. That is, use as a check figure the remainder of each number after dividing it by five:

Problem	Check
7	2
× 8	3
56	2 × 3 is 6. Reduce by dividing by 5 and showing only the remainder: 1

Five-remainder of answer: 1. Right.

Note that in none of these check figures do we count the answer to any division by the "base" of our check figure. It is only the remainder we watch—because the remainders must stay in order through the calculations. If the remainders do not check out, we know the answer is wrong.

As a general exercise in number sense, try "casting out" the sevens in the next example. Your check figure in each case is now the *remainder* after dividing by *seven,* and you use the check figures just as you would use digit sums:

$$\begin{array}{r} 58,792 \\ -49,296 \\ \hline \end{array}$$

Cover up the explanation below with your pad while you do this problem (from left to right, canceling in the answer) and then check your results by dividing each number by seven and using only the remainder as your check figure. Handle the check figures just as you would digit sums.

Here is the working:

Problem	Check
5 8 , 7 9 2	6
− 4 9 , 2 9 6	2
$\not{1}$ 9 , $\not{5}$ $\not{0}$ 6	4

Check figure of answer: 4

The one weak point of casting out any single-digit number for checking purposes is that any one digit in your answer that happens to wrong by the exact size of the digit you are casting out will not be caught. "Casting out" (or dividing by) a two-digit number is by nature more accurate. The easiest two-digit number to cast out—which also depends on every digit in the number when casting it out, unlike ten for example—is 11. There are three different ways to test divisibility by 11, or to determine the remainder after a division by 11 to use as a check figure. None of them is quite as simple as adding up the digits (which casts out nines), but with a little practice it goes quite fast.

Dividing By Eleven

In your work with numbers in the past, you may have learned to recognize numbers exactly divisible by eleven because of the pattern they form.

All two-digit numbers divisible by eleven, for instance, are paired digits—from 11 through 99.

For two-digit numbers, then, you can quickly get the elevens-remainder by subtracting from the number (mentally) the next lower number with paired digits. Here are some examples:

$$\begin{array}{r} 46 \\ -\ 44 \\ \hline 2 \end{array} \qquad \begin{array}{r} 83 \\ -\ 77 \\ \hline 6 \end{array} \qquad \begin{array}{r} 25 \\ -\ 22 \\ \hline 3 \end{array} \qquad \begin{array}{r} 64 \\ -\ 55 \\ \hline 9 \end{array} \qquad \begin{array}{r} 92 \\ -\ 88 \\ \hline 4 \end{array}$$

Note with special care that next-to-last example. When you cast out elevens, nine is no longer "0." Nine is "0" only for digit-sum purposes. Both nine and ten are check figures you will use when casting out elevens. When you cast out elevens, *eleven* becomes 0. Since you are using remainders as check figures, within the check system the number you cast out becomes 0.

The check figure of 88, when you cast out 11's, is 0. The check figure of 98 is ten. The check figure of 97 is nine. Don't forget and call it 0.

Numbers from 100 to 999 also form a particular pattern when exactly divisible by 11. The two "outside" digits of any three-digit number will (when added) equal the "middle" digit or else exceed it by 11—if the number is divisible by 11. In other words, a three-digit number is exactly divisible by 11 if the sum of the first and third digits equals the middle digit or else exceeds it by 11.

Here are some examples:

8 9 1	1 9 8	2 7 5	3 6 3	2 2 0
2 0 9	3 0 8	7 0 4	3 1 9	6 3 8
8 1 4	1 5 4	4 2 9	1 1 0	5 0 6

At this point, the pattern becomes more of a figuring job and less an obvious shape you can "scan" as you glance at the number. The above examples, particularly if you test them out by dividing with 11 and watching *why* the patterns form as they do, is an excellent exercise in number sense. Just as important, however, they lead to two general rules for determining 11's remainders.

Numbers divisible by 11 continue to form patterns, but more complicated ones, as the number of digits goes above three. The patterns, however, are the reasons why the rules work. Try the first rule on the above numbers to gain some feeling of why it works.

Odd and Even Digits

A quick way to extract a check figure based on division by 11 is to subtract the total of all the digits in even places (starting from the right) from the total of all the digits in odd places.

In the first example above, the only even-placed digit is 9. (Even, of course, means divisible by two.) The first and third, or odd, digits (starting from the right) are 1 and 8. These total 9. 9 from 9 is 0. The 11's remainder is 0.

In deciding "odd" and "even" places, you always start from the *right*. This is the only place in the entire book where you are permitted to read a number from right to left, but you have to for this purpose.

In the last example above, the only even-placed digit is 0. The total of the two odd-placed digits (5 and 6) is 11. Perhaps you can guess that, since 11 is 0 for 11's-remainder purposes, you are in effect subtracting 0 from 0—or if the middle (even) digit were 2, you would be subtracting 0 from 2.

If you have any trouble remembering whether "even" or "odd" comes first—is to be subtracted from the other—just recall that E (for even) appears in the alphabet before O (for odd). In professional memory-expert circles this is called a mnemonic key. After a few days' disuse, such a key can be very useful.

Here is how this technique works with a few numbers you already can "feel":

2 3 4 6 3 0 8 1 5 4 4 2 9

In order, here is the working:

The even-placed digit (counting from the right) in 23 is 2. 2 from 3 is 1. This is the 11's remainder.

In the number 46, you subtract 4 from 6 and find the check figure 2. Test this against dividing 46 by 11 and finding the remainder.

For 308, the even-placed digit is 0. Subtract this from the sum of the odd-placed digits (3 plus 8) or 11. The result is 11. For 11's-remainder purposes, this is 0.

Do the last two on your own.

Now one complication creeps in. Sometimes, you will find that the total of the even-placed digits is greater than the total of the odd-placed digits—and not always by an exact 11, which we consider to be 0. Consider:

6 9 1

The only even-placed digit is 9. The total of the odd-placed digits is 7. You cannot subtract.

But, as you might suspect in this system, you can *add 11* to that 7 and then subtract. 7 plus 11 is 18. 18 minus 9 is 9.

The rule is this: When the total of your even-placed digits is smaller than the total of your odd-placed digits, add 11 to the total of the odd-placed digits and then subtract.

This method works on numbers of any length. In general, it is most useful for numbers of three, four, and five digits. Above that, another method will become more useful. First, however, reinforce your understanding of the even-from-odd method by trying it on the following numbers:

7 9 1 2,6 4 8 5 4 0 8,6 2 3

The 11's remainders of these four numbers are, in order, 10, 8, 1, and 10.

The even-from-odd technique is useful primarily for numbers in which you can spot the even numbers and hold their total in your mind while adding the odd numbers, then (after adding 11 if necessary) subtract. The optimum size for rapid "scanning" (after some practice) is four or five digits. For longer numbers, still a third alternative becomes most useful.

Continuous Subtraction

For any number, no matter how many digits it contains, there is a technique for finding the 11's remainder in one continuous process from left to right. It is not (alas) quite as much of a snap as adding up digit sums, but it is as simple as we can make it. Once you really learn the technique, you will find it amazingly swift.

The method is to subtract the first digit from the second, this result from the third, this result from the fourth, and so on through the very end of the number. If any succeeding digit is too small to be subtracted from, add 11 and then subtract.

Notice how it works on a simple example:

1 3 4

Start by subtracting 1 from 3. Answer, 2. Now subtract this answer from the next digit: 2 from 4. Answer, 2. Test the correctness of this 11's check figure by finding the remainder by the even-from-odd method: 3 from the sum of 1 and 4 is also 2.

Try the continuous subtraction technique on this number:

1 3 5 7 9

Working from the left, the process goes: 1 from 3 is 2, from 5 is 3, from 7 is 4, from 9 is 5. 11's remainder, 5. Verify it, if you wish, by subtracting the total of the even-placed digits from the total of the odd-placed digits: 7 plus 3 is ten. 9 plus 5 plus 1 is 15. 10 from 15 is 5.

So far, continuous subtraction seems almost as easy as digit sums. Now, however, try it on the same number reversed:

9 7 5 3 1

To start with, you cannot subtract 9 from 7. First you must add 11 to the 7, then subtract: 9 from 18 is 9. This 9, in turn, cannot be subtracted from 5. It can, however, be subtracted from 5 plus 11: 9 from 16 is 7. Once more, you have to add 11 to the 3 before you can subtract: 7 from 14 is 7.

Adding 11 to the final 1, you find that 7 from 12 is 5. The 11's remainder is 5.

This number is an extreme. On the average, you have to adjust with an extra 11 in about half of the digits, not all of them. A more typical process would go like this:

4 6 1 7 9 8

Here is how it goes: 4 from 6 is 2, from 12 (1 plus 11) is 10, from 18 (7 plus 11) is 8, from 9 is 1, from 8 is 7. 11's remainder, 7.

Take special care to go through any zeros at the end of the number. Zeros after a decimal point do not count (unless followed by another digit), but zeros before a decimal must be included in your calculation. For instance:

1 0 0

You can "feel" what the 11's remainder of this is by mentally subtracting the next-lower two-digit number with paired digits: 99 from 100 is 1. Continuous subtraction, for demonstration, would go like this: 1 from 11 (0 plus 11) is 10, from 11 (0 plus 11) is 1.

Use of Complements

If you have learned your complements thoroughly, you will find that they can speed up this process. You subtract, of course, by adding the complement of the number to be subtracted to the number from which you are subtracting—if the number to be subtracted is larger than the other.

You can make a routine of this for continuous subtraction, with the extra little kicker that you add one *extra 1* each time you use a complement. This gives the same result as adding 11.

Try this technique on this number:

8 4 2 5 3

Complement-kicker subtraction goes like this: Complement of 8 (2) plus 4 *plus 1* is 7; complement (3) plus 2 *plus 1* is 6; complement (4) plus 5 *plus 1* is 10; (no complement) plus 3 *plus 1* is 4. 11's remainder, 4.

Checking Addition

Except that you extract your check figures in a different fashion, proving your answers with 11's works precisely the same way as checking with digit sums. Find your 11's remainders, do to them whatever you did to the numbers, and the result must equal the 11's remainder of the correct answer.

When adding, you add the check figures, reduce if need be by casting out the 11's of your total (you can no longer reduce by adding the digits, remember; that is for digit sums only) until you have a final check figure of 10 or less. This is equal to the 11's remainder of the answer.

Follow the checking of this problem step by step:

Problem	Check
4 6 9	7
2 3 1	0
9 8 6	7
7 9 4	2
2 2 6 0	1 6
2 2	Reduce (1 from 6) to 5
2 4 8 0	

11's remainder of answer: 2 from 4 is 2 from 8 is 6 from 11 (0 plus 11) is 5; *or* 10 (8 plus 2) from 15 (4 plus 0 plus 11 to adjust) is 5.

Try this one on your pad:

```
 6 3 8
 1 4 7
 2 6 9
 _____
```

Work out the answer and check it with 11's before comparing your results with this explanation:

The check figure of 638 is 0; of 147 is 4; of 269 is 5.

The total of these is 9. The correct answer is 1054, which has a check figure of 9: 1 from 11 (for the 0) is 10, from 16 is 6, from 15 is 9. Or the even-placed digits 5 and 1 total 6, from 4 plus 0 plus 11 (to adjust) is 6 from 15, or 9.

Checking Subtraction

In subtraction, just as in using digit sums, you subtract your check figures to see if the result equals the check figure of your answer. If the check figure of the larger number is smaller than the check figure of the smaller number, add 11 to it before subtracting. If you prefer, add the check figures of the answer and smaller number; this must equal the check figure of the larger number.

Problem	Check
6 4 9 5 8 2 1	2
− 5 6 3 6 7 8 9	4
1 8 0 9 1 4 2	You cannot subtract 4 from 2. Add 11 to 2, and subtract 4 from 13: 9.
(8 5 9 0 3 2)	

Check figure of answer: 9.

Try this one on your pad before looking at the answer and its proof:

```
  3 5 4 6 7 8
− 2 4 6 3 2 6
```

Remember to work from left to right and cancel in the answer.

The 11's remainder of the larger number is 5, of the smaller number is 3. 3 from 5 is 2. The check figure of the correct answer, 108352, is 2. Right.

Checking Multiplication

You prove your multiplication answer by multiplying the check figures of the numbers you multiplied to see if the result—reduced by casting out 11's—equals the check figure of your answer.

Problem

$$
\begin{array}{rrrrrrr}
 & 9 & 8 & 4 & 7 & & \\
 & & \times & 6 & 2 & & \\
\hline
 & 5 & 8 & 0 & 8 & 2 & \\
 & & \bar{1} & 9 & 6 & 9 & 4 \\
\hline
 & 5 & 0 & 9 & 4 & 1 & 4 \\
(& \bar{6} & 1 & \bar{0} & \bar{5} & 1 & 4 \;)
\end{array}
$$

Check

$$
\begin{array}{c}
2 \\
\underline{7}
\end{array}
$$

2×7 is 14, which reduces (1 from 4) to 3.

Check figure of answer: 3. Try it yourself.

Now carry one through on your own:

$$
\begin{array}{r}
7\;3\;5 \\
\times\;4\;8 \\
\hline
\end{array}
$$

Cover the answer and its proof with your pad until you have finished.

The 11's remainder of 735 is 9. The check figure of 48 is 4. 9×4 is 36, which reduces (3 from 6) to 3. The correct answer is 35280, and has a check figure of 3.

Checking Division

You recall that in checking division with digit sums, you could not divide the digit sums even though you had divided the numbers. This is inherent in all check figures because (with the two remainders we use as check figures) either 9 or 11 is "0."

Just as in checking with digit sums, you check with 11's by multiplying the check figure of the answer by the check figure of the divider—adding the check figure of the remainder, if any—and seeing if this equals the check figure of the number divided.

Here is an example:

Problem	Check	
3 7 3 (last 3 underlined)	Answer	0
5 8 / 2 1 6 9 4	Divider	3
1̸ 4 2	Remainder	2
1̸ 2 3	Number divided	2
1̸ 6 0		
1̸		

The check figures work like this: 11's remainder of answer (0) times remainder of divider (3) is 0, plus check figure of remainder (2) is 2. The 11's remainder of the number divided is 2. Everything checks out.

Try this one, working out the solution in shorthand division and checking it by casting out 11's:

3 7 / 9 1 7 6

Cover the explanation with your pad until you have finished.

The answer is 248, which has an 11's check figure of 6. There is no remainder. The check figure of the divider is 4. 6 × 4 is 24, which reduces (2 from 4) to 2. The check figure of the number divided is also 2. Perfect.

Duplicate Proofs

Several times, we have mentioned the advisability in critically important cases of double-checking. Unlike the traditional double-check of doing the problem over twice in opposite directions, the use of both 9's and 11's gives an absolute, unquestioned proof of accuracy—completely divorced from the human possibility of multiplying 4 × 8 and getting 28 three times in a row.

Here is one final example of division, the trickiest both to solve and to check, worked out and proved in both ways:

Problem	9's Check		11's Check
2 6 2	Answer	3	9
4 6 / 1 2 5 6 8	Divider	1	2
1 3 3			
1 6 0	Remainder	1	10
2 4	Number divided	4	6
1 5 6			
1 0			

Proof with 9's: 3 × 1 is 3, plus 1 is 4. Check figure of number divided is 4.

Proof with 11's: 9 × 2 is 18, which reduces (1 from 8) to 7. 7 plus 10 is 17, which reduces (1 from 7) to 6. Check figure of number divided is 6.

12

HOW TO USE SHORT CUTS

IN THIS chapter we shift gears entirely, and learn how to build on our simplified arithmetic a different, but very useful, system of conversion.

This second section does not ignore the first. Indeed, the swift and confident working of these new principles depends very much on a smooth and automatic handling of your basic number combinations—the practical, fast, fully integrated system of speed arithmetic based on the phenomenal simplicity of the modern Japanese abacus.

Before going on to the short cuts, let your mind range back over what you have learned.

In complement addition, you have learned to add from left to right. You have discarded the twenty hardest digit combinations by using a complement instead of adding over ten. You have learned to record and forget tens as you go along, picking them up at the end of each column.

In complement subtraction, you have learned never to subtract a larger digit from a smaller, but instead to use a complement. You have learned to avoid the cumbersome, confusing system of "borrowing," by accomplishing the same thing in the answer itself. This lets you work from left to right.

In no-carry multiplication, you have learned to work from left to right too. You also now have the technique for producing the answer swiftly and easily without carrying—without juggling a number of digits in your mind at the same time or stopping to make notes of them as you work.

In all three operations, your answer takes form naturally in your head or on the paper—just as it takes form naturally on the abacus.

In shorthand division, you have learned to combine the techniques of no-carry multiplication and complement subtraction with the European "shorthand" method of long division, shorten it still further in several respects, and get your answer with half the pencil work needed in traditional long division. You have also learned a simple but effective secret for producing your next answer digit in a flash by dividing with only the first digit of the divider, raised by one, and adjusting it later if you need to. This, also, has been adapted from abacus theory.

This entire system is a remarkably fast and easy approach to numbers, once you fully master it. But it also lends itself so beautifully to speeding up still further the better standard short cuts that you can compound your new handling of figures by carrying the "abacus" system into short cuts.

The approaches of the two methods are very different.

Our simplified arithmetic is an integrated system. Standard short cuts are not. They consist of a variety of tricks, mainly in converting problems to simpler forms, that apply to a wide variety of problems but not to all of them. Further, they have never before been assembled into anything resembling a unified whole.

I hope to show before this book is finished that the more useful of the standard short cuts can be learned in relation to each other, so that you can reach for the most effective in any particular case. More than on any other single foundation, this "integrated" approach to the short cuts will rest on the firm base of your number sense.

Perhaps this is the appropriate time to explode some of the fables about number short cuts. Valuable short cuts there are, but not one of them will enable you to multiply 38,657 by 49,956 in ten seconds in your head, without effort, and with perfect accuracy. History has been made by a number of mathematical geniuses, but a thorough sifting of the evidence exposes the single secret possessed by every one of them. It is a secret, alas, beyond most of us.

Jedidiah Buxton, the illiterate son of an English schoolteacher, was able to calculate entirely in his head the problem: "Multiply two times two 140 times, then assume that this answer is in quarter-pennies and reduce it to pounds, shillings, and pence." His answer was in 39 digits. Oddly enough, however, Buxton did not know very much about arithmetic. His methods were almost unbelievably crude. Instead of multiplying by 300, for instance, he would multiply by 5, then 20 (in effect multiplying by 100, which he could have done by adding two zeros), then multiply the result by 3. Buxton's one secret was the basic secret of all the "mental calculators" of history—a staggeringly complete memory for figures. This enabled him to handle immense numbers in his head, even doing calculations that took days or weeks, remembering every digit as if returning to a mental blackboard when he resumed work on a problem.

Thomas Fuller was a slave who showed no signs of his unusual gift until he was 70, a little after the American Revolution. Then he gave demonstrations on the order of finding the number of seconds in 70 years, 17 days, in just a minute and a half. Zerah Colburn, the son of a Vermont farmer, started at eight. His father took him on an exhibition tour, and he brought a skeptical academic audience in England to the verge of tears by giving the 16th power of 8 (8 used as a multiplier 16 times) faster than the answer could be written; the answer, incidentally, is 281,474,976,710,656.

The point to these stories is simply this: the secret possessed by every mental calculator is nothing more or less than a prodigious memory for numbers. Some of them became real mathematicians; others never learned to apply their mental oddity to anything more serious than number stunts to impress paid audiences. In some cases, their actual understanding of numbers, as such, was ludicrous.

For most of us who have trouble remembering what to "carry" or whether or not we "borrowed" because we had to deal with another number in the meantime, the most fruitful approach is to strip down our methods to the simplest, fastest, and easiest techniques.

In a number of cases, however, there is a way to "see through" a seemingly complex problem—see through the apparantly unrelated figures and reduce them to a simple, sensible relationship that we can almost recognize at a glance. It cannot produce an exact answer to every problem; instead, it picks and chooses, from the many problems we must solve, the one-half to three-quarters, roughly, that can be solved in less time if we convert them to other forms first.

The classic short cuts are really nothing more or less than methods of conversion—conversion from one form into another form entirely, where the relationships can more easily be seen or solved.

For instance, if faced with the problem $15 \overline{)45}$, you might or might not recognize the answer at once as 3. If not, and if you were trained and alert to conversion possibilities, you would note that 15 doubled is 30—a far simpler divider. Being knowledgeable about short-cut methods, you would double the 15 and the 45 as well (to keep the relationship identical) and see the problem as $30 \overline{)90}$. This can be nothing but 3.

Most short cuts are basically as simple as that, which is an example of "proportionate change."

Another example we have mentioned before is the short-cut method of figuring a 15% tip. The 15, you will note, is the same as the 15 in the last paragraph. But in this case most experts use another short cut. Start, for example, with a meal check for $4.00. 15 is exactly $\frac{1}{10}$ of a hundred plus $\frac{1}{2}$ of that tenth. The 15% tip on $4.00 then is 40¢ ($\frac{1}{10}$) plus 20¢ ($\frac{1}{12}$ the tenth), or 60¢.

This method is called "breakdown."

Could you use the method we first mentioned, proportionate change, to find the tip? Yes, though multiplication works in a different fashion from division. In division you double both numbers to keep the relationship the same. In multiplication, to keep the relationship the same you must cut one of them in half if you double the other. $\frac{1}{2} \times 2$ is 1—so the problem will have the same answer.

To use proportionate change on this tip, you would double the 15 to 30, and cut the $4.00 in half to get $2.00. 30% of $2.00 is (3 × 2) 60¢ again.

Which is the better short cut? Neither. Each one fits certain combinations of numbers better than the other does, and it is helpful to know both so you can select the easier of the two for any one case.

This is the central fact about short cuts. There are literally hundreds of short cuts, from the quick method of squaring a number ending in 5 (when did you last have to square a number ending in 5?) to multiplying by 11 in one operation (a little more useful, but still pretty specialized). Of the many available, only four types of short cut are really applicable to enough problems to be worth the trouble of learning, unless for the reward of knowing a number oddity to impress people.

These four short cuts have certain similarities and certain differences. Learn them well, learn how to recognize which is most valuable in any one case, and you can add to your already advanced handling of basic numbers the extra advantage of frequent "overleaps" in lightning calculation.

Dig out your pad again, or better yet get a fresh one, and prepare to enter another fascinating aspect of numbers.

13

BREAKDOWN

IT IS hardly likely that you would ever multiply a number by ten by putting down the number with ten under it, and multiplying out digit by digit like this:

```
  4 6 3
  × 1 0
  -----
  4 6 3
    0 0 0
  -------
  4 6 3 0
```

Instead, you know that in order to multiply any number by ten you simply add a zero. If the number has a decimal point in it, you move the decimal point to the right instead of adding the zero.

984 × 10 is 9840.

653.92 × 10 is 6539.2.

Elementary as this is, the principle is basic to many of the short cuts in number work. In many cases, we can save time by multiplying or dividing a number by ten, a hundred, or even a thousand before even beginning work.

In division, of course, you remove a zero (or move the decimal point one place to the left) in order to divide by ten.

2390 divided by 10 is 239.

718.64 divided by 10 is 71.864.

In order to avoid any possible confusion, make sure you understand that any whole number is *presumed* to have a decimal after it. We shall get more deeply into the subject in the chapter on decimals, but for the moment let's point out that 75 can be considered to be 75.00. Then, if we divide by 10, we move that presumed decimal one place to the left. 75 divided by 10 is 7.5.

Each digit, you remember, increases tenfold in value as it moves one place to the left. So to multiply by a hundred, we add *two* zeros (ten times ten), or move the decimal point two places to the right.

984 × 100 is 98400.

653.92 × 100 is 65392.

When dividing by a hundred, we also move the decimal point two places—to the left.

984 divided by 100 is 9.84.

653.92 divided by 100 is 6.5392. We would most likely round it off to 6.54.

Undoubtedly none of this is new to you. It is merely a refresher. But the refresher is important, because the more easily and automatically you can *think* this multiplication or division by ten or a hundred, the more quickly and confidently you will handle the short cuts that involve such division or multiplication as a basic part.

Our second step into the breakdown short cut is through another obvious technique that may well be second nature to you already.

In dealing with many numbers, you probably know already how to multiply by 9 in the "round off and adjust" method. Rather than multiplying by 9, you multiply by 10—and subtract 1.

Compare the two methods:

Usual way	Breakdown way
6 5 9	6 5 9 0
× 9	− 6 5 9
5 $\underline{8}$ 3 1	$\not{6}$ 9 $\not{4}$ 1

The working in these two examples is not dramatically different, but they are cited to illustrate a point and to lead into more sophisticated examples. Once again, for the sake of your number sense, try to "feel" the identity of the two expressions above of precisely the same situation.

Just as you probably already knew this special dodge in handling 9, it is likely that you have used in the past the same sort of approach in handling numbers very near 100.

If you have to multiply 238 by 99, surely you would not bother to set up the whole problem and multiply it out line by line. You just subtract one 238 from a hundred 238's, as this comparison demonstrates:

Usual way

$$\begin{array}{r} 238 \\ \times 99 \\ \hline 1042 \\ 1042 \\ \hline 23562 \end{array}$$

Breakdown way

$$\begin{array}{r} 23800 \\ -238 \\ \hline 23\not6\not72 \end{array}$$

If you were required to multiply by 101, on the other hand, you would simply add one 238 to a hundred 238's. This, after a very moderate amount of practice, you easily do in your head. After a few tries, you should be able to "see" the answer as 24038.

Not very often is your work as extra-simple as multiplying by 99 or 101. But the principle works in a surprising variety of cases, and is the "round off and adjust" special subdivision of our first general short cut: breakdown.

The over-all rule for breakdown is this: break one of your numbers down into *two* easier-to-handle numbers.

Thus we broke 9 down into 10 and 1.

We broke 99 down into 100 and 1.

We can also—here is where the method becomes far more generally useful—break 45 into 50 less 5. 5, you note, is exactly $\frac{1}{10}$ of 50. Or we break 44 down into 40 plus 4—the 4 being exactly $\frac{1}{10}$ of 40.

Stop for a moment and try the first example:

Usual way	Breakdown way
6 2 9	3 1 4 5 0 (50 × 629)
× 4 5	− 3 1 4 5 (1/10 above)
2 4 1 6	3̸ 8 3 1̸ 5
3 1 4 5	
2 8 2 0 5	

It is especially helpful when you can break down a number into two parts of which one is an even fraction of the other, such as 50 and 5. You cannot always do this, of course, which is why we also use other short cuts.

The exact breakdown may well depend on the relationship between the numbers to be multiplied. In some cases one breakdown will make sense, in other cases quite a different breakdown.

Note how it varies in these two cases:

1 2	6 2
× 1 8	× 1 8

Which breakdown of 18 might you use in the first example? The number 18 can be broken into 12 plus 6 (½ of 12), into 9 plus 9 (two equal parts), into 20 minus 2 (1/10 of 20).

For the first example, the most convenient breakdown of 18 might well be 12 plus 6—because most of us have dealt enough in grosses to know almost by instinct that 12 × 12 is 144. So 18 × 12 is 144 plus ½ of 144 (72)—which we can see as 216.

For the second example, however, most of us could not quickly "see" the answer to 62 × 12. If we break down 18 into 9 plus 9, we can quickly multiply 62 × 9 and then add the answer to itself. Furthermore, we can multiply by 9 using 10 minus 1. This is two-step breakdown. Complex as it may seem at first glance, a very quick and simple way of solving this example would be to handle it as "620 minus 62—doubled."

If we chose the third breakdown, our number work would be surprisingly similar to that involved in the second. 20 minus

2 is identical to 10 minus 1, doubled; only the order of operation is changed.

The advantages of this sort of breakdown show up more dramatically, of course, in longer numbers. Try one of the last two breakdowns of 18 on the following problem. Use your pad and pencil:

```
4 9 3 6 2
    × 1 8
---------
```

Let us choose the (10 minus 1) doubled breakdown for 18 in this case. Here is how the work should look:

```
  4 9 3 6 2 0   (10 × 49362)
−   4 9 3 6 2   (minus one 49362)
-------------
  4 5̸ 4 3̸ 6̸ 8
  8 8 8 4 0 6   (doubled)
        ‾ ‾
```

As with many of the demonstrations, the short-cut nature of the method is not as striking at first sight as you will find it in actual practice. Often you will use almost as many figures, and as many operations. But you are using basically simpler combinations: multiplying by 10 instead of by 9; subtracting instead of doing another digit-by-digit multiplication; doubling instead of adding two lines.

Just for comparison, here is how the 20 minus 2 breakdown for 18 works in the same problem:

```
  9 8 7 2 4 0   (20 × 49362)
−   9 8 7 2 4   (1/10 above)
-------------
  8̸ 8̸ 8̸ 5 1̸ 6
```

In this case, it is presumed that you can jot down twice any figure at sight, and add a 0 at the end to get the effect of multiplying by 20.

There is virtually no limit to the breakdowns you can find. You can break down a number into two parts that add up to the original number (such as 12 plus 6 in 18, or 10 plus 5 in 15) or two parts of which you subtract one from the other to get the number (such as 100 minus 1 for 99, 20 minus 2 for 18, 60 minus 6 for 54).

How would you break down 81? Depending on the number you needed to multiply, you could make it 80 plus the original number, or 90 minus $\frac{1}{10}$ of the *product* (since 90 minus 9 is 81, and 9 is $\frac{1}{10}$ of 90).

Your proportions need not always be $\frac{1}{10}$. They can be $\frac{1}{2}$, $\frac{1}{3}$, $\frac{1}{4}$, or any other convenient fraction. The key is to find a *convenient* fraction, or there is no sense in using the breakdown method.

See if you can recognize convenient breakdowns for these numbers:

39 26 77 63 125 720

Of 39, we can make 40 minus 1. Of 26, we would make 25 plus 1. In another short cut, incidentally, you will find a far easier way to use a number such as 25 than by multiplying by 2 and then 5 and adding. 77 is obviously 70 plus $\frac{1}{10}$ of the product, while 63 is 70 minus $\frac{1}{10}$ of the product. We can tackle 125 in several ways; for this use, we can consider it 100 plus $\frac{1}{4}$ of the product. 720 is 800 minus $\frac{1}{10}$ the product.

In the choice of short-cut methods, and in the best use of each, you have great flexibility. There is no substitute for number sense here, for it is in finding the relationships that your key to method selection lies. There are so many variations, so many slightly different approaches, that it is up to you to select the fastest and easiest in each case.

Try these problems on your pad, finding an appropriate breakdown for each:

$$\begin{array}{r} 895 \\ \times 15 \\ \hline \end{array} \qquad \begin{array}{r} 473 \\ \times 38 \\ \hline \end{array} \qquad \begin{array}{r} 682 \\ \times 27 \\ \hline \end{array}$$

We have already covered the most convenient breakdown for 15. It was inserted here to remind you of the repetitive character of many useful breakdowns. In multiplying 895 by 10 and adding half the product, you would think simply "8950, plus 4475, is 13425."

In dealing with 473 × 38, you run into another fraction in your breakdown. 38 is 2 less than 40. 2, in turn, is $\frac{1}{20}$ of 40. So, to multiply by 38, you can multiply by 40 and subtract

$\frac{1}{20}$ of the product. $\frac{1}{20}$ is just $\frac{1}{2}$ of $\frac{1}{10}$. You do it like this:

```
 1 8 9 2 0   (40 × 473)
   - 9 4 6   (1/20 of product)
 ---------
 1 8̸ 0̸ 8̸ 4
```

Note two instructive points about this example:

First, you can (and should) jot down the answer to 40 times 473 from left to right without copying the original number. It is simply 4 × 473, digit by digit in the no-carry method, plus one zero.

Second, you can (and should) jot down from left to right the division of 18920 by 20 without any strain. You simply divide by 2 and start one place to the right when you put down the answer, which also divides automatically by 10. The combination results in a division by 20.

The third breakdown, 682 × 27, breaks down the 27 into 30 minus $\frac{1}{10}$ the product. Here again, you multiply 682 by 3 as you jot down the result and add one zero to make the multiplication by 30 instead of by 3. Under it you write the same digits one place to the right, without the zero, which automatically divides by 10, and then subtract:

```
 2 0 4 6 0   (30 × 682)
 - 2 0 4 6   (1/10 above)
 ---------
 2̸ 8 4 2̸ 4
```

Short cuts are a variety of methods, not a single system. There are many problems to which you can find short cuts, others to which you cannot, without doing more work than simplified arithmetic would involve. It comes down to recognizing the short cut that makes sense in a flash, because if you brood for more than an instant or two on whether or not to use a short cut at all, in that time your new systems of basic arithmetic could have finished most of the problem.

Try recognizing breakdown possibilities in these numbers:

50 45 24 33 54 63 82

Some of these begin to pioneer new breakdown possibilities that we have mentioned but not yet fully demonstrated. Yet your own good number sense should show you interesting ways in each case.

50, for instance, is exactly half of 100. It is entirely up to you whether you find it easier and quicker to multiply by 5 and add a zero, or to add two zeros and divide by 2. Simple as it may seem, this is a perfectly valid short cut.

45 has been mentioned before, as 50 times the number, less $1/10$ of the product. If you have not noticed it before, 45 is also 30 times the number, plus $1/2$ the product. Which is better? Neither. It depends on the relationships of the numbers with which you are working, and on your own preferences.

24 is a new one. 24 is 20 times the number (double it and add a zero), plus $2/10$ of the product. Jotting down $2/10$ is simple: double the product, but start one place to the right. This divides by 10 and multiplies by 2 at the same time. If this seems at all obscure, follow the working in this example:

Usual way	Breakdown way
3 7 8	7 5 6 0 (20 × 378)
× 2 4	1 4 1 2 ($2/10$ × product)
0 7 5 6	8 0 7 2
1 4 1 2	
8 0 7 2	

This breakdown example emphasizes the value we put at the beginning of this chapter on being able to handle multiplication and division by 10, 100, etc., without hesitation or strain. This is the key to handling breakdowns such as the one above of doubling a number and then doubling the result —but multiplying by 10 in the first case and dividing by 10 in the second.

Our next number is 33. This is obviously 30, plus $1/10$ of the product.

63 is based on the opposite principle. 63 can easily be broken down into 70, minus $1/10$ of the product.

82 is a bit different. We will break 82 down into 80, plus $1/40$ of the product. This is not difficult to handle. You simply divide the product by 4, but start writing your answer one extra place to the right. This divides by 4 and by 10 all at once —dividing by 40. See how it works:

Usual way	Breakdown way

9 4 2
× 8 2
7 5 3 6
  1 8 8 4
7 6 1 4 4
	7 5 3 6 0 (80 × 942)
	  1 8 8 4 (¼0 above)
	7 6 1 4 4

In a division such as the one above, you have an automatic running check on your accuracy because the division must come out even. If it does not, you know you have made a mistake. This is because two whole numbers, when multiplied, must give a whole-number answer. So if your division has a remainder, you are warned to recheck it.

There is no clear-cut advantage in this particular problem to breaking down 82 in the fashion we did, rather than into 80 plus twice the original number (which is merely a simpler expression of what our regular multiplication does). In one case you divide by 4 and 10; in the other you double. If the product of 80 × 5555 were part of the problem, however, it would be very tempting to divide the first product of 44440 by 4 and 10 to get the second line. Once again, which breakdown is best depends on how the numbers relate to each other.

By and large, the major value of breakdown is in permitting you to use easier-to-handle operations and digits. Breaking down 78 into 80 minus twice the other number, for instance, lets you substitute a simple doubling for a multiplication by 8 at the second step.

Breakdown—like any short-cut technique—is valuable to you only as you learn to handle it easily and well. Do not dismiss it out of hand if your first reading of a particular problem leaves you more baffled than enlightened, but on the other hand do not force yourself to use a particular short cut that after a few tries does not spring into your mind naturally and obviously. The purpose is to save work, not make it.

Longer Numbers

The easiest numbers to break down are usually those with two digits. But this does not mean that much longer numbers cannot also be broken down, frequently with dramatic results.

Take the multiplier 297, for instance. The nearest one-digit number that can form the base of your breakdown is 300. The difference between 297 and 300 happens to be a very convenient 1/100 of the product.

Note the same feature in the numbers 396—495—594. For each of them, you can substitute a multiplication by the next even hundred and subtract 1/100 of the product, instead of multiplying through by three digits and then adding all three lines.

In reverse, the same short cut is possible with 303—which you have probably used in the past without special instruction. There is no need to multiply twice by 3; merely copy the first product again, two places to the right, and add.

Now that you have learned to add or subtract 1/10 or 1/100, 2/10 or 2/100, and so on, the possible range becomes considerably larger. You might handle 306, for instance, by doubling the first product two places to the right, rather than multiplying by 6.

As the breakdowns become more complex, so does the saving of time in using them. When you break down a three-digit number into two one-digit parts, you save a full digit in your work while at the same time performing a basically simpler operation.

For instance, consider the multiplier 784. There is no simple relationship between 700 and 84, so you do not break it down that way. 784 is just 16 less than 800, however, and 16 is exactly 2/100 of 800. So, instead of multiplying digit by digit by 784, we can multiply by 800 and subtract 2/100 of the product. Now the short cuts become visibly dramatic:

Usual way	Breakdown way
3 4 6	2 7 6 8 0 0 (8 × 346, plus 00)
× 7 8 4	− 5 5 3 6 (2/100 above)
2 3 2 2	2 7 1 3 7 4
2 7 6 8	
1 3 8 4	
2 6 0 1 6 4	

In finding the first line of the working figures for the breakdown example, you do not copy the problem itself. You should be able to work without copying the problem with any single-digit multiplier if you have been working conscientiously on your no-carry multiplication. The second line of working figures, of course, is merely the first line doubled—two places to the right.

Since you are beginning to find it more and more natural to multiply by any single digit, you can extend your breakdowns into any number of tenths or hundredths. The number 558 might, at first sight, not show any exciting breakdown possibilities. 58 bears no reasonably simple proportion to 500. 558 is 42 less than 600. The key is to look at the 42 and the 6 in 600, and note—6 × 7 is 42. So you can multiply any number by 558 by first multiplying with 600 and then subtracting $\frac{7}{100}$—which you do by multiplying the first product by 7 and putting down the answer two places to the right.

Choosing Multipliers

In all of our demonstrations so far, we have broken down the bottom number of the problem—the one normally considered to be the multiplier. Except when problems are set up for us in this fashion, there is of course no real "multiplier" and "number multiplied." In actual business or personal life, we simply need to multiply two numbers together, and it is up to us to decide which we will treat as the multiplier.

The reason this fact is worth special attention is that you can break down *either* number of a multiplication. As you start a particular problem, glance at both numbers for breakdown possibilities. The one you break down becomes your multiplier.

For example, you might face the problem 69 × 58. A quick look at 69 shows you that it can be broken down into 70, minus 1. 58 can be broken down, but not nearly as easily. So pick 69 as your multiplier.

Mixed through the various examples so far have been

two different methods of breakdown. One is the special case called "rounding off and adjusting," in which you choose a convenient round number and then add or subtract the other number or a simple multiple of it to adjust. 69 is an example of this. So might be 68, since it is easier to multiply by 70 and then subtract twice the other number than it is to multiply it first by 6 and then by 8 and then add.

The second method is rounding off and adjusting by a fraction of the product of your first multiplication, rather than by the other number. For 63, you multiply by 70 and subtract $\frac{1}{10}$ of the product. For 392, you multiply by 400 and subtract $\frac{2}{10}$ of the product.

This difference should be crystal clear. In the first case, your difference is adjusted in terms of the number multiplied. In the second case, your difference is adjusted in terms of the product of your first multiplication.

Here, in order to make the difference very specific, is the same number broken down in each way:

Other-Number Adjustment	First-Product Adjustment
48—50 minus 2 times the other number	48—40 plus $\frac{2}{10}$ the product

Which of the two breakdowns is better? Once again, neither. It depends on the other number and on the methods you yourself find easiest to handle. Either breakdown, you note, permits you to substitute a simple doubling for a multiplication by 8.

You can push the breakdown technique to impressive extremes. The nearest convenient one-digit multiplier may not be the next even ten or hundred at all; it may be two or more away. 1860, for instance, can be broken down so that you multiply by 60 and add 30 times the product. 328 can be-become: multiply by 8 and add 40 times the result.

This field of sophisticated breakdowns is fascinating, but it is too involved to be treated fully here. If you enjoy the idea, you can doodle for hours and find a breakdown for almost any number you may try. As genuinely useful short cuts, however, the more abstruse applications are questionable. You would spend more time breaking down your multiplier than the whole problem would take in simplified arithmetic. Number sense, again, is the real key. If you cannot "see" a relationship at one or two glances, then the short cut is not a real short cut for you.

The most useful ground rules for the two types of breakdowns are these:

ONE: If you round off one of the numbers to be multiplied, can you add or subtract the other number to adjust few enough times to be easier than the full multiplication?

TWO: If you round off one of the numbers to be multiplied, can you add or subtract a simple enough fraction of the first product to be easier than the full multiplication?

If the answer to either of these questions is yes, then breakdown can save you work and time in solving the problem. If the answer seems to be no, then another short cut may be in order.

Answering these two questions rapidly is the way to break down problems quickly and easily. See how many sensible breakdowns you can find in these multipliers:

58	936	52
197	78	392
72	180	9
288	29	54
36	192	18
69	784	27
63	88	90
270	89	582
686	240	39
26	714	3984

Each of these numbers can be broken down in a way that

will save you work. Some of them save you quite a bit of work; others let you add or subtract instead of multiplying by a high digit; still others reduce the number of lines of working figures. Try them on your pad before you check your reactions against the proposed breakdowns that follow.

In some cases, more than one breakdown is possible. We will give only the one that seems simplest and most generally useful. Since the breakdowns are of both types, we will use the shorthand N to mean that adjustment is in terms of the other number, and P to mean that adjustment is in terms of the product of the first multiplication.

For instance, our breakdown for the first number—58—is given as $60 - 2N$. This means you multiply by 60 and then subtract the other number, doubled. The breakdown for 72 is given as $80 - \frac{1}{10}P$, which means you multiply the other number by 80 and then subtract $\frac{1}{10}$ of the product.

Here are the breakdowns:

58 ($60 - 2N$)
197 ($200 - 3N$)
72 ($80 - \frac{1}{10}P$)
288 ($300 - \frac{4}{100}P$)
36 ($30 + \frac{2}{10}P$)
69 ($70 - N$)
63 ($70 - \frac{1}{10}P$)
270 ($300 - \frac{1}{10}P$)
686 ($700 - \frac{2}{100}P$)
26 ($20 + \frac{3}{10}P$)
936 ($900 + \frac{4}{100}P$)
78 ($80 - 2N$)
180 ($200 - \frac{1}{10}P$)
29 ($30 - N$)
192 ($200 - \frac{4}{100}P$)
784 ($800 - \frac{2}{100}P$)
88 ($80 + \frac{1}{10}P$)
89 ($90 - N$)
240 ($200 + \frac{2}{10}P$)
714 ($700 + \frac{2}{100}P$)
52 ($50 + 2N$)
392 ($400 - \frac{2}{100}P$)
9 ($10 - N$)
54 ($60 - \frac{1}{10}P$)
18 ($20 - \frac{1}{10}P$)
27 ($30 - \frac{1}{10}P$)
90 ($100 - 10N$)
582 ($600 - \frac{3}{100}P$)
39 ($40 - N$)
3984 ($4000 - \frac{4}{1000}P$)

Two or three special notes are in order. The idea of "breaking down" 9 may seem peculiar. Yet it is possible, should you choose to use it; and you may well prefer to subtract a number from the same number (with an added 0) rather than to multiply through the entire number by 9.

The same comment applies to 90, of course. It is precisely the same breakdown, with one more 0 on both numbers.

Breaking down the number 26 into 20 plus $\frac{3}{10}$ the product does not, in one sense, save any steps. The point here is that it offers you the choice of multiplying the other number by 6, or the first line of working figures by 3 (starting one place to the right). Other factors being equal, it is usually easier to multiply by the smaller of two digits—in this case, by 3 rather than by 6. So while breaking down 26 is not a short cut in the sense of saving steps, it does simplify the operation.

Breakdown in Subtraction

Ninety per cent of the value of breakdown is in multiplication. There is no easy way to use it in division, and it does not really save any time in addition. In subtraction, however, breakdown can sometimes speed up a problem if the relationship of the numbers is within a certain range.

The technique in subtraction is to raise the smaller number to the next-higher even number, then add the same amount to the larger number. This converts the problem into a form in which you can see the answer at a glance.

Suppose you need to subtract 64¢ from 98¢. Using the breakdown technique, you add 6 to 64 to make it an even 70. You adjust by adding 6 to 98 too, which then becomes $1.04. Subtracting 70 from 104 is a sight job. In subtracting 297 from 465, you add 3 to 297 to make it an even 300, and adjust by adding 3 to 465 to make 468. The answer, 168, is automatic.

The main application of this method is in adjusting numbers that fail by merely a digit or two of reaching the next even number. If the adjustment is much more than this, complement subtraction will be both easier and faster.

For such special cases, however, breakdown can be useful. Here is one example:

Usual way	Breakdown way
6 8 4 5 9	6 8 4 5 9 + 7 is 6 8 4 6 6
− 4 9 9 9 3	− 4 9 9 9 3 + 7 is 5 0 0 0 0
~~2~~ ~~9~~ ~~5~~ 6 6	Answer (at sight) 1 8 4 6 6

While you will not find such examples in your work every day, they do come up once in a while and this little trick is well worth keeping in mind.

14
ALIQUOTS

THIS fascinating and useful technique of conversion suffers under a traditional and foreign-sounding name. "Aliquot" means, simply, an exact fraction. The word is derived from a Latin word meaning some, or several. It is usually used as an adjective (aliquot parts, meaning exact parts), but since it is also a noun we will save words.

The key word in the definition is *exact*. 8 is an aliquot of 16, because it is contained within 16 exactly twice and leaves no remainder.

Since we count by the decimal system, based on ten, the aliquots of most use to us in short-cut mathematics are aliquots of ten, a hundred, a thousand, and so on. Incidentally, the word is pronounced ali-kwut.

We all think of 25¢ or "a quarter" as completely interchangeable, without giving it a second thought. We have dealt in quarter-dollars so much that we know by instinct that 25¢ is one quarter of 100¢. The special usefulness of this and many other aliquots (for 25 is indeed an aliquot of 100) may or may not have been brought to your attention.

For instance, you can multiply by 25 by adding two zeros to the other number and then dividing by 4:

```
Usual way              Aliquot way

6 8 2                      1 7 0 5 0
×2 5                   4 / 6 8 2 0 0
1 3 6 4                    2
  3 4 1 0
1 6 0 5 0
```

The value of aliquots is not restricted to the number 25 (or its equivalents 250, 2500, 2.5, .25, and so on). Half of 25 is 12½, and 12½ is a number we meet surprisingly often. It is exactly ⅛ of 100. The same aliquot shows up as 125 (⅛ of 1000), as 1.25 (⅛ of 10), as .125 (⅛ of 1).

You might soon need to multiply 965 by 12.5. Which of these two ways looks easier?

```
Usual way              Aliquot way

  9 6 5                    1 2 0 6 2.5
× 1 2.5                8 / 9 6 5 0 0
  0 9 6 5                  1     2 4
    1 9 3 0
      4 8 2 5
  0 0 9 6 2.5
```

The number 5 is also an aliquot, of course. It may be a tossup whether you would prefer to multiply by 5, or add a 0 and divide by 2. It depends on which you find easier. A very similar approach was suggested for 50 in the chapter on breakdown, incidentally; this illustrates the overlapping nature of some of the features of the different short-cut methods.

There are only 11 useful exact aliquots in the decimal system, but they are number combinations that show up very often. In addition, there are a number of *approximate* aliquots which can prove useful in estimating—such as 33 for ⅓ of 100—but be sure to remember that they are not real aliquots at all.

Here are the 11 aliquots. In order to avoid decimals, we will show them as aliquots of 1,000. Adding zeros, or moving decimal points to the left, can make these same numbers prove to be aliquots of anything from 1 to any number of million you wish.

Exact Aliquots

based on 1,000

125	$\frac{1}{8}$	400	$\frac{2}{5}$	750	$\frac{3}{4}$
200	$\frac{1}{5}$	500	$\frac{1}{2}$	800	$\frac{4}{5}$
250	$\frac{1}{4}$	600	$\frac{3}{5}$	875	$\frac{7}{8}$
375	$\frac{3}{8}$	625	$\frac{5}{8}$		

All the 16th's, by the way, are exact four-digit aliquots, except $\frac{1}{16}$, but since the fraction is in two digits (16) their utility for short-cut arithmetic becomes somewhat remote. $\frac{1}{16}$ of 10,000 is precisely 625, while $\frac{3}{16}$ of 100,000 is 1875. $\frac{2}{16}$—naturally—is the same as $\frac{1}{8}$, which appears in the table above.

Even aliquots with top and bottom digits (such as $\frac{3}{8}$) can save work, because the number 375 for which $\frac{3}{8}$ is the aliquot contains three digits. In order to multiply by 375 in the aliquot way, you first divide by 8 (after adding three 0's to the other number, since 375 is $\frac{3}{8}$ of 1000) and then multiply the result by 3. Although you first divide and then multiply, this is still a little simpler than multiplying through with each of three digits and then adding the three lines of partial products.

Here is a comparison of the two methods:

Usual way

```
   8 4 3
 × 3 7 5
 -------
   2 5 2 9
     5 8 0 1
       4 2 1 5
   -----------
   2 0 5 1 2 5
```

Aliquot way

```
         1 0 5 3 7 5 × 3 is 0 3 1 5 1 2 5
  8 / 8 4 3 0 0 0
          3 6 4
```

Do one on your own now. Cover the explanation that follows with your pad until you have solved this problem with an aliquot:

```
2 4 7 4 7
    × 2 5
---------
```

This is a very simple one, but you may be surprised at how much work an aliquot can save you even in a case like this.

Multiplying 24747 by 25 is, naturally, precisely the same as dividing 100 times 24747 by 4. So that is what we do. Our answer is

```
        6 1 8 6 7 5
  4 / 2 4 7 4 7 0 0
          3 2 3 2
```

Work out the answer to the problem in the traditional way and look at the two workings, side by side. The difference is quite dramatic.

Try one more, before moving on to other applications of the aliquot short cut. The following problem can be solved by using two aliquots, one for each stage of the solution. See if you can decipher this:

```
 3 6 5 4
×6 2 5
-------
```

As always, cover the explanation with your pad until you have finished.

625 is an aliquot of 1,000, being ⅝ of it. Instead of multiplying by 625, then, we can divide 1,000 times 2654 by 8 and then multiply the result by 5. First, let us show the straight comparison:

Usual way

```
  3 6 5 4
× 6 2 5
---------
1 1 9 2 4
  0 7 3 0 8
    1 8 2 7 0
-------------
2 1 7 3 6 5 0
```

Aliquot way

```
        4 5 6 7 5 0 × 5 is 2,283,750
    8 / 3 6 5 4 0 0 0
        4 5 6 4
```

The second-stage aliquot solution here can come in multiplying the 456,750 × 5. If you find it easier to add a 0 and divide by 2 instead of multiplying by 5, you can easily set up this step into the answer of the first. Your working then looks like this:

```
        2 2 8 3 7 5 0
    2 / 4 5 6 7 5 0 0
  8 / 3 6 5 4 0 0 0
      4 5 6 4
```

Even in so complex a solution as this, the aliquot method obviously involves fewer working figures. Compare it with the standard solution once more.

Special Aliquots

The fact that many of our measuring systems are non-decimal (not based on ten) gives them different sets of aliquots. ¼ of ten, for instance, is 0.25. But the gallon is based on eight, not ten (two pints in a quart, four quarts to a gallon), so in terms of pints ¼ of a gallon is 2.

This gives us an occasional and interesting interplay between regular ten-base aliquots and gallons, feet, yards, hours, and other non-decimal measurements.

We can see at a glance that one pint is precisely 0.125 gallon. If we need to know how many pints are in 0.8750 gallon, we find that the 8 in the fraction form of the aliquot 875 ($\frac{7}{8}$) is wiped out by the conversion from decimal to pints-gallons, and we are left with an even 7 pints.

Inches to feet is a little tougher, since $\frac{1}{12}$ does not have a precise decimal equivalent. In other terms, $\frac{1}{12}$ is not an aliquot of the ten-base system, because its decimal equivalent is .0833+, with 3's going on forever because it never becomes exact. It is very close, however, so except for complete scientific accuracy you will find it accurate enough.

To find the number of inches in 0.9166 feet, then, you would note that the approximate fraction of .9166 is $\frac{11}{12}$. In converting from decimal to duo-decimals (dozens), the 12 gets dropped and you have 11 inches.

Here are the most frequently used approximate aliquots. Remember that these are not true aliquots, because they are not precise, but they are close enough for a great deal of your number work.

APPROXIMATE ALIQUOTS

based on 1,000

83	$\frac{1}{12}$	583	$\frac{7}{12}$
167	$\frac{1}{6}$	667	$\frac{2}{3}$
333	$\frac{1}{3}$	833	$\frac{5}{6}$
417	$\frac{5}{12}$	917	$\frac{11}{12}$

It is interesting to note that all the approximate aliquots are based on thirds and multiples of thirds—sixths and twelfths. This is inherent in the ten-based (decimal) system.

An extra bonus in the use of aliquots to bridge the difference between a ten-base system and an eight-, twelve-, or other-base system in weights and measures is that becoming aware of the aliquot equivalents is one of the best exercises you can give your number sense.

Try it once yourself. Using aliquots, figure out the number of pints in 375 gallons.

It should not take long. 375 is an exact aliquot, being

⅜ of 1,000. Since there are 8 pints in a gallon, there would be 8,000 pints in 1,000 gallons. The 8's cancel out, and you are left with 3,000 pints.

How many months in 83 years, for a quick guess? 83 is an approximate aliquot, about ⅚ of 100. ⅚ is of course the same as $^{10}/_{12}$, and there are 12 months in a year. So the 12's cancel out, and we have about 10 times 100—or a thousand months. Actually it is 996, so we are .4 of 1% off.

Dividing with Aliquots

Unlike breakdowns, aliquots are just as valuable in dividing as in multiplying. When you divide with an aliquot, you simply reverse the rule for multiplying.

In multiplying, you multiply by the fractional form of your aliquot. In dividing, you divide by the fraction.

In multiplying, you add enough zeros to the other number to make the aliquot stay in proportion. 50 is ½—*of 100*—so to multiply by 50 with a division of 2, you first add two zeros to the other number.

In dividing, you *subtract* as many zeros as you need to. Usually, you must use a decimal point.

Let us start with one of the simpler aliquots. Here is how you use the aliquot 25 for dividing:

Usual way	Aliquot way
3 8 4	9 6.0 0
2 5 / 9 6 0 0	× 4
2 1	3 8 4
1 1 (first 1 struck out)	

Note that we subtracted two zeros from the number divided by using the decimal point. We subtracted two zeros because 25 is ¼ of 100. If we had been dividing by 2.5, we would have subtracted one zero because 2.5 is ¼ of 10. Dividing by 250 would require us to subtract three zeros.

The reason you almost always have to use a decimal point to subtract zeros when dividing with an aliquot is that division often does not come out even. The example above was a simplified introduction. If the number to be divided were 9643, of course, then we know simply by inspection that there would be a remainder because subtracting two zeros (by moving the decimal to the left) from 9643 gives us 96.43, and those two digits to the right of the decimal must be multiplied too.

Try a longer division yourself. Cover the answer with your pad until you have finished:

$$125 \overline{)73984}$$

The proper aliquot form to use for 125 is ⅛. Since 125 is ⅛ of 1,000, we subtract three zeros from the number divided. Here is how the problem is set up:

$$\begin{array}{r} 73.984 \\ \times 8 \\ \hline 5\underline{8}1.872 \end{array}$$

If you feel ambitious, you might try dividing 73984 by 125 in the usual way to see if you get the same answer—and to compare the amount of work involved.

Since you multiply from left to right, you may not have to finish this multiplication all the way through. Carry it to the accuracy you need and then stop. If you need only the nearest tenth, work it out through the 7 and round off your answer to 591.9.

In aliquots with two digits, you again reverse the multiplication process. In multiplying, you divide by the bottom figure of the fraction (the 8 in ⅝) and then multiply by the top digit. In dividing, you multiply by the bottom digit and then divide by the top. This, naturally, is equivalent to division by the fraction.

Suppose we go through the following problem with an aliquot solution:

$$87.5 \overline{)40530}$$

First, determine the aliquot. 87.5 is ⅞ of 100. Since we are using a fraction of 100, we subtract two zeros from the other number and start by multiplying it with the bottom of the fraction:

```
 4 0 5.3 0
       × 8
----------
3 2 4 2.4
```

Now—and you would not bother to rewrite the result in actual practice—you divide by the top of the fraction:

```
      4 6 3.2
 7 / 3 2 4 2.4
      4 2 1
```

This is obviously much easier than dividing, even in shorthand long division, by a three-digit number.

Turn to a clean page of your work pad now and tackle this problem with an aliquot solution:

```
7 5 / 2 9 7 0 0
```

Cover the answer with the pad.

The fraction for the aliquot 75 is ¾ of 100. First we subtract two zeros (we can simply omit them here, since there are two zeros) and multiply by the bottom of the fraction:

```
 2 9 7
   × 4
-------
0 1 8 8
```

Now divide this product by the top of the fraction. In practice you would do it at sight:

```
       3 9 6
 3 / 1 1 8 8
       2 1
```

That is all there is to it. Without dividing by anything more difficult than the single digit 3, you know that 29700 divided by 75 is 396.

Reversing Aliquots

If you will turn back for a moment to the table of exact aliquots, you will note that several of them are really simpler in their decimal form than they are in their fractional form.

A later chapter will cover fractions and decimals. If this special application of their interchangeability in terms of aliquots is at all confusing, it might be a good idea to refresh your memory with that chapter first.

The fraction $\frac{4}{5}$, for instance, has the aliquot form .8. The decimal form of $\frac{2}{5}$ is .4.

This fact makes possible a reverse short cut whenever you must deal in fractions that are more simply expressed in decimals. Rather than suffer through the fraction, use the simpler form.

One example should illustrate this sufficiently. Consider this problem:

$$\frac{3}{5} \times 2\ 8\ 7$$

This problem would traditionally be solved by multiplying 3 × 287 and then dividing the product by 5. But it is far, far easier to multiply 287 × .6:

$$\begin{array}{r} 2\ 8\ 7 \\ \times .6 \\ \hline 1\ \underline{6}\ 2.2 \end{array}$$

The important lesson in this reverse-aliquot approach is that no *single* method is always best. The point is to learn awareness of the many different ways of accomplishing the same result, and to be on the lookout for the easiest and quickest in each particular case.

Sometimes it will be breakdown. Sometimes it will be the use of an aliquot. And sometimes it will be the use of factors—quite a different short cut.

15
FACTORS

MOST of us remember, from our school days, the word "factors." Chances are you have not encountered the word or the process since. Instead of considering them merely an exercise for students, however, we will show how they can short-cut many problems in multiplication and division.

A factor means, basically, a maker or doer. The word has many applications in English. In mathematics it means one of two or more numbers which, multiplied together, produce the number in question.

6 has two factors: 2 and 3. 2 and 3 are factors of 6 because 2×3 gives 6.

Almost three-quarters of all numbers are factorable. That is, they can be broken down into two or more other numbers which, multiplied together, produce the number you started with. Of the first hundred numbers (from 1 to 100) only 26 are prime numbers. Prime numbers are those that cannot be factored.

1 and 2 are both prime numbers, because they cannot be factored. It is true that 1×1 is 1, but we do not consider 1 to be a legitimate factor. It would not be of any use to us in short-cut mathematics, in any event. 3 is also prime. But 4 can be factored into 2 and 2, because 2×2 is 4.

Before going into the ways of factoring numbers, let us

show the exciting possibilities in their use. They are a powerful short cut because they can save major steps in multiplying and dividing.

To multiply by a factorable number, multiply first by one of its factors and then multiply the result by the other. Where is the short cut? Watch:

```
 6 8 4
× 5 6
──────
```

In order to use factors, we first find a number that can be factored. Even though in real-life situations you will look at both parts of a multiplication rather than arbitrarily decide that one of them is the multiplier, it is usually quicker to consider the shorter of the two numbers the multiplier.

In this case, 56 is the multiplier. Can it be factored? Can you think of two other numbers that, multiplied together, produce 56? Your knowledge of the multiplication tables should snap the factors 7 and 8 into your mind.

The factor short cut in multiplying any number by 56, then, is to multiply first by 7, then the result of that multiplication by 8. Compare the two ways:

Usual way	Factor way
6 8 4	6 8 4
× 5 6	× 7
3 4 2 0	4 7 8 8
3 0 0 4	× 8
3 8 3 0 4	3 7 2 0 4

The two examples may *look* about equally time-consuming. But note than in the usual way you multiply first by 5, then by 6, then add the two products to get your final answer. In the factor method you still multiply by two digits—7 and then 8—but you never add any partial products at all. You save roughly one-third the work.

Let's do another before you try one on your own. Check this problem for factor possibilities:

```
 7 9 3
×2 8
```

If you have "seen" the factors of 28 at a glance, let us compare methods again:

Usual way	Factor way
7 9 3	2 9 3
× 2 8	× 4
1 5 8 6	2 1 7 2
5 3 4 4	× 7
2 1 1 0 4	2 1 1 0 4

Once more, we managed to skip entirely the step of adding two lines of partial products. Multiply by 4, then by 7, and you have the final answer.

Try this one by yourself. Cover the answer with your pad as you work:

```
 2 6 7
×3 6
```

This problem has two different, equally correct factor solutions. You can factor 36 into 6 and 6, or into 4 and 9. Any series of accurate factors will produce the same result. Compare the working with these two sets:

2 6 7	2 6 7
× 6	× 4
1 5 0 2	0 0 6 8
× 6	× 9
0 9 5 1 2	0 9 5 1 2

Incidentally, there are several other factors of 36. You could use 2 and 18, or 3 and 12. But each of these sets involves two-digit factors. These are useful in longer problems, but it always pays to seek the simplest solution. For simplicity, the choice between two digits and one digit is plain.

How to Factor

We set aside until now the question of factoring itself, so we could show how it works in multiplication. With this specific encouragement, we will get down to the process of factoring before going on to division.

For numbers up to 100, you should be able to recognize factors pretty much at a glance. The most useful factors are single digits, and these can carry you up to 81. Some two-digit numbers can be factored only with three factors or factors of which one has two digits (we will get into the handling of these later), but 74 out of the first 100 numbers can be factored.

Just for a taste, go through the numbers 40 to 49 to see what the possibilities are:

40— 5 × 8
10 × 4
20 × 2

41—prime

42— 6 × 7
2 × 21
3 × 14

43—prime

44— 4 × 11
2 × 22

45— 5 × 9

46— 2 × 23

47—prime

48— 6 × 8
3 × 16
4 × 12
2 × 24

49— 7 × 7

Just because you can factor seven out of these ten numbers (several of them in more than one way) you should not think that you should always use factors. Rule number one for all short cuts remains: look for every possibility, then do it the easiest way. Sometimes you will use factors, and sometimes you will pass them up even if they would be possible, because another short cut happens to be easier or because your new simplified arithmetic is easiest of all in this particular case.

Sharpen your factor-eye by trying the numbers from 30 to 40. Jot down all the possibilities you see on your pad before checking with the following table.

Here are the factors for numbers in the 30's:

30— 5×6
10×3

31—prime

32— 4×8
2×16

33— 3×11

34— 2×17

35— 5×7

36— 4×9
6×6
2×18
3×12

37—prime

38— 2×19

39— 3×13

So much for the numbers up to 100. Some of those that can be factored are easier to use straight than with factors (38, for instance), but such two-digit factors lead us into higher numbers where they can be very valuable indeed.

It becomes more difficult to recognize factors at sight when we get above 100. Yet factors are even more useful for numbers going into several digits, because they often become dramatic short cuts for bigger numbers.

How would you know, for instance, that 261 can be factored into 9 and 29? Or 536 into 8 and 67?

There are very definite keys developed over the years that show almost at a glance whether a number can be factored with a single digit as one of the factors.

You already know one of them from your work in casting out nines. The digit sum of 261 above is 0; this means that the remainder after dividing by 9 is 0. Obviously, then, 9 must be a factor of 261.

Casting out elevens, too, is merely testing a number for even divisibility by eleven. If there is no elevens-remainder, then eleven is a factor of that number.

Here, in numerical order, are the keys to determining the divisibility of any number by 2 through 12—except for 7. There is a key for 7, but it is so hideously complicated that it is in no sense a short cut.

Key for Divisibility

2 If it is even (the last digit can be divided by 2).

3 If the digit sum is divisible by 3. Just cast out the 9's, and if the remainder is 3 or 6 the number is exactly divisible by 3. (If 9 is a factor, it can obviously be broken down further into 3 × 3, but there isn't much point in doing so because this would raise the other factor in the same proportion.)

4 If the last two digits are divisible by 4. 536 above has 4 as a factor, because 36 can be evenly divided by 4. Two 0's as the last two digits also make it divisible by 4.

5 If the last digit is 0 or 5.

6 If it is divisible by both 2 *and* 3, as outlined above.

7 Too complex a key to be useful here.

8 If the last *three* digits are divisible by 8. There is a simpler approach to this in actual working, however, since the other factor (when found) will often show you how to increase 4 to 8. Wait and see in the examples to come.

9 If the digit sum is 0.

10 If it ends in 0, of course.

11 If the 11's-remainder is 0. Check back with the chapter on the back-up check.

12 If it is divisible by both 3 and 4, as outlined above.

Try these keys on the two examples mentioned earlier. How can you tell that 9 is a factor of 261? Because the digit sum is 0. How do you know what the other factor is? Simply by dividing with 9, which is simple with a one-digit divider. 9 goes into 261 exactly 29 times, so the factors of 261 are 9 and 29.

Now take 536. You know at sight that 4 is a factor, because the last two digits (36) are divisible by 4. The next step is to find the other factor by dividing 536 by 4. This gives you 134 as the other factor. BUT—since 134 is even, you can double the 4 (to 8) and cut 134 in half, to 67. This is the simpler approach mentioned in the key table to divisibility by 8. If you start with 4 and find that the other factor is even, double the 4 to 8 and cut the other factor in half.

As a general rule, it is helpful to use the *largest* one-digit factor you can, because this cuts the other factor down to the smallest possible size. For 536, it is obviously easier to deal with 8 and 67 than with 4 and 134.

For a bit of practice, factor these numbers. Use your pad to cover the answers:

114	603	392
345	159	139
486	546	243

Warning: one of these numbers is prime, but only one. All the others can be factored.

Here is how we factor all but the next-to-last of the above numbers:

114: It is even, so 2 is a factor. The digit sum is 6, so 3 is also a factor. If both 2 and 3 are factors, then we know 6 is a factor. The factors of 114 are 6 and 19.

345: This ends in 5, so 5 is a factor. 5 into 345 gives you 69 as the other factor. You will note that 3 is also a factor, but this would give you the set 3 and 115. 5 and 69 is an easier pair.

486: Digit sum, 0. The factors are 9 and 54.

603: Digit sum, 0. The factors are 9 and 67.

159: Digit sum, 6. 3 is a factor. It is not even, so 2 is not a factor, and if 2 is not a factor then neither is 6. 3 is the largest single-digit factor, and the other (by division at sight) is 53.

546: The digit sum is 6, and it is also even. Both 2 and 3 are factors, so the highest factor is 6. 546 is produced by the factors 6 and 91.

392: 2 is a factor (the number is even), but 3 is not because the digit sum is 5. The last two digits, however, are divisible by 4. So we start with the factors 4 and 98. Since 98 is even, we simplify matters by doubling the 4 and halving the 98, and get the factors 8 and 49.

139: This is a prime number. It has no factors. Try all the keys.

243: The digit sum is 0. The factors are 9 and 27.

How to Factor Factors

So far, you have learned to multiply by two one-digit factors. In order to multiply by 56, you multiplied first by 7 and then by 8. In our last number above, however, is it really simpler to multiply by 9 and then by 27 rather than by 243?

It might not seem to be at first thought, but it really is. When you multiply by 243 you have three lines of partial products to add. Using the factors, you have only two (from the two-digit factor).

Often, however, you can factor the factors. You recognize 27 as 3 × 9. So the factors of 243 are 9, 9, and 3.

Extend the factor solution now to include *three* factors. Instead of multiplying by 243, multiply first by 9. Multiply this result by 9. Multiply this result, in turn, by 3. The answer will be correct, as this comparison of all three solutions shows:

```
 Usual way          Two factors         Three factors

   7 8 4            7 8 4               7 8 4
 × 2 4 3              × 9                 × 9
 -------            -------             -------
 1 5 6 8            6 0 5 6             6 0 5 6
   2 1 3 6          -                   -
   -                  × 2 7                   × 9
     2 3 5 2        ---------           -----------
 -----------        1 4 1 1 2           6 3 4 0 4
 1 8 9 4 1 2          4 9 3 9 2             -
     - -            -----------                 × 3
                    1 8 0 4 1 2         -----------
                      -   -             1 8 0 5 1 2
                                          -
```

examples. In the usual way you multiply by each of three digits,

Do not be deceived by the lines of type occupied by these

then add three lines of partial products. With two factors, you again multiply by three digits, but add only two lines of partial products. With three factors (each of one digit) you multiply still by three digits—but never do any adding at all.

See if you can factor the multiplier in the following example into three single-digit factors, and work out the problem on your pad before reading on:

$$\begin{array}{r} 8\ 4\ 7 \\ \times\ \underline{3\ 3\ 6} \end{array}$$

There are several ways you might have factored 336. You could have started with 6 (digit sum 3, number is even) or 4 (36 is divisible by 4). Suppose you started with 4. This would produce the factors 4 and 84. Since 84 is even, you would immediately change the factors to 8 and 42. Since you recognize 42 as the product of 6 and 7, you have the three factors 8, 6, and 7.

One technique we have not yet mentioned, which is frequently quickest, is to start multiplying with your largest factor. It is usually easier for most of us to multiply quickly by smaller digits than by larger ones, and the working figures you multiply get larger at each step.

Our three-factor solution to this problem then goes like this:

$$\begin{array}{l} 8\ 4\ 7 \\ \underline{\quad\ \times 8} \\ 6\ 7\ 7\ 6 \\ \underline{\qquad\ \ \times 7} \\ 4\ \underline{6}\ \underline{3}\ 3\ 2 \\ \underline{\qquad\qquad \times 6} \\ 2\ 8\ 4\ 5\ 9\ 2 \end{array}$$

Does this answer check out? Try nines- or elevens-remainders on it and see.

Dividing with Factors

Factors are as useful for division as they are for multiplication. Division is by nature the most difficult of the four basic processes for most of us, and you may like the application of factors to division most of all because they frequently permit us to divide with single digits rather than more complex numbers.

Division is just the reverse of multiplication, so the use of factors in division is just the reverse of their use in multiplication. The technique is to factor the divider if you can, then divide by each of the factors. Each division, of course, is into the result of the last division rather than into the original number divided.

Watch how it works in this case:

```
63 / 137214
```

The factors of 63 are 7 and 9. Divide first by 7:

```
       19602
  7 / 137214
```

Now divide the *result* by the other factor, 9:

```
       2178
  9 / 19602
```

Compare this solution to the usual method of working:

```
          1177
   63 / 137214
        1̸74
          11
          1̸5̸9
             50
             1̸63
```

Which way, even at first glance, *looks* easier? The faster nature of the factor short cut becomes even more dramatic if you divide the second factor into the result of the first division without bothering to rewrite, like this:

```
        2178
   9 / 19602
  7 / 137214
```

The work should be clear. You started at the bottom and worked up, setting up the second division into the answer of the first. It is merely a condensed picture of the two stages shown separately in the first explanation, and is the way you would actually do it in practice—assuming you did not merely jot down the answer to the first division without bothering to rewrite the problem.

Which Factor to Use First

In multiplication, we start with the largest factor and work down. In division, we *usually* start with the smallest factor and work up—for precisely the same reason, in reverse. The easiest digits to divide by are usually the smallest, and our division stages get smaller as we progress. So for the earlier divisions into longer numbers, it is most often easier to start with the smallest factors.

Watch out for special cases, however, particularly in problems with remainders—which so many in division have. In these cases you may be able to get through the first division without a remainder if you handle it properly. This simplifies things.

In general, match the factor used first to the number divided. If one factor is odd and the other even, start by dividing with the odd factor if the number divided is odd and by dividing with the even factor if the number divided is even.

Here is an example that illustrates this:

$$28\ \overline{)\,10619}$$

The first step is to factor the divider into 4 and 7. Second, note that this division *cannot* come out even; it must have a remainder. You know this because even into odd can never produce an even answer (although odd into even can). So, in this case, we start with the odd factor rather than the even one:

$$\begin{array}{r} 1517 \\ 7\ \overline{)\,10619} \end{array}$$

If we started with the even factor, here is what the first step would look like:

```
      2 6 5 4.7 5
4 / 1 0 6 1 9
```

Obviously, the other approach is easier to begin with. Dividing this result by the second factor, now, we produce the final answer:

```
      3 7 9.2 5
4 / 1 5 1 7
```

This illustration does not bother to put the decimal point and zeros into the second number divided because you do not need to either. Just keep mentally "bringing down" zeros after the decimal point.

If you try dividing 2654.75 by 7, you will get the same final answer. But it is more work. You had a remainder on the first division by 4, so you have to divide through two remainders instead of just one.

Matching odd and even will not always avoid this, but it often will. When you cannot avoid a remainder in the first result, by the way, carry it only to as many decimal places as you will need in the final answer. There is no point to dividing on and on with a remainder that may never come out even.

One other key to watch for in picking your first-division factor is to see if either of the factors of your divider is also a factor of the number divided. If it is, you know that the first division by this factor must come out even.

Now try one division by factors on your own:

```
7 2 / 3 2 9 0 4
```

Cover the answer with your pad until you have finished.

The factors of 72 are recognizable at sight: 8 and 9. Since the number divided is even, we will start with the even factor, then divide that result by the other factor. The illustration will be in condensed style. See if your working agrees with this:

```
          4 5 7
     9 / 4 1 1 3
  8 / 3 2 9 0 4
```

Division with Three Factors

Just as it often pays to use three factors in multiplying, so you can use three factors in division to speed up and simplify your work.

In multiplying, you use each of the three (or more) factors in turn to avoid adding lines of partial products. In dividing, you use each of the factors in turn to avoid the extra complications of dividing with a number of two or more digits —where you can. When one of the factors has to be of two or more digits, you will still find the division simpler than dividing with a still longer number.

Suppose we factor this problem:

```
5 6 7 / 4 7 9 8 5 2 1
```

No matter how we tackle it, this is admittedly one of the divisions most of us hate to face.

First, see if the divider can be factored. 567 has a digit sum of 0, so we know at once that 9 is a factor. 9 into 567 (without writing down the problem, of course) gives us the other factor, 63. 63 we recognize as 7 × 9. So 567 has three one-digit factors: 9, 9, 7.

Since all the factors are odd, you may as well start with the smallest. Stack your working as we did before, and the factor solution looks like this:

```
            8 4 6 3
       9 / 7 6 1 6 7
    9 / 6 8 5 5 0 3
 7 / 4 7 9 8 5 2 1
```

What factoring really accomplished in this case, as you can see, was to reduce the solution to three divisions of single digits each, rather than one division by a number of three digits.

It is time now to try a three-factor division yourself. Cover the answer with your pad until you have finished this problem:

$$224\,/\overline{172032}$$

The first step is to see if the divider can be factored. The digit sum is 8, so it is not divisible by 9 or 3. It is even, so it is divisible by 2, but moreover the last two digits are divisible by 4, so 4 is a factor. We start by factoring it into 4 and 56. Since 56 is even, we double the 4 and cut the 56 in half: factors, 8 and 28.

We prefer to work with single-digit factors if we can, so we further factor the 28 to 4 and 7. All the factors we need for 224, then, are 4, 7, and 8.

Since the number divided is even, let's start with the 4 and work up:

```
              7 6 8
        8 / 6 1 4 4
      7 / 4 3 0 0 8
    4 / 1 7 2 0 3 2
```

That's all there is to it. That rather fearsome division problem is reduced, thanks to factoring, to three quick and simple single-digit divisions.

Sometimes you can factor a number into one one-digit factor and one two-digit factor, but cannot further factor the two-digit factor at all. You may still save time by using these two factors, though, just as you can in multiplication. Dividing by a two-digit number is so much easier than dividing by a three-digit number (even in the shorthand method) that it will probably pay you to use the factors—especially since you have already gone to the trouble of factoring the divider.

Every time you use factors, you will become fonder of them. They are the third major area of short-cut conversions, following naturally after breakdown and aliquots.

Now we will take up the fourth type of conversion.

16

PROPORTIONATE CHANGE

THE fourth generally useful type of short cut has no traditional name. Because the phrase most accurately describes what we do, we will call it proportionate change.

The technique is simply that: proportionate change. You change one number of a problem into a simpler form in *any* way you wish (double it, triple it, cut it to one-third, or whatever) but change the other number *in proportion* so the essential relationship remains the same.

For instance:

$$45 \overline{)180} \quad \text{becomes} \quad 90 \overline{)360}$$

The conversion here should be quite obvious. One glance at the problem shows us that 45 can be converted into a one-digit (plus 0) number by doubling. So we double it, without hesitation, to the more easily handled 90.

The *proportionate* part of the rule is simpler for division than it is for multiplication. In division, you do to the number divided precisely the same thing you did to the divider. In multiplication, you do to the other number exactly the opposite of whatever you did to the first number.

In the example above, you double 45 to 90. To keep the proportion, you now double the 180 to 360. The answer, simply by inspection, is obviously 4—9 into 36.

Try one yourself:

$$35 \overline{)210}$$

Start by examining the divider for any simple change that will convert it to a one-digit number—plus a 0 if necessary. Doubling 35 changes it to 70. Do the same thing to the number divided, which changes it to 420. Most of us would find it rather difficult to "see" the answer to 35 $\overline{)210}$, but the answer to 70 $\overline{)420}$ should be a matter of reading *ea* and *sy* as "easy."

There is another way of handling the proportions, incidentally, and this is to change the answer rather than the number divided. In division, you can change the divider—divide into the original number divided—, then change the answer in the same way you changed the divider.

Our example above now becomes 70 $\overline{)210} \times 2$. "See" the answer; it is the same one we got before. According to the numbers and the change involved, this is sometimes easier.

Doubling is only one of the proportionate changes you can use. You can triple, quadruple, or multiply by any number you choose. Or you can cut in half, in thirds, in quarters, as you will. Remember that in division and multiplication you cannot add or subtract, however; the change must be in the nature of a multiplication or division. And remember to compensate in the other number or in the answer.

In its simplest terms, here is an illustration of a problem clearly calling for one specific proportionate change:

$$33\tfrac{1}{3} \overline{)5600}$$

The simplest change here is to triple the divider. This gets rid of the fraction and reduces the whole divider to one working digit to boot. You divide by 100—and multiply the number divided or the answer by 3 to compensate.

The general rule becomes obvious from this example and the former ones. Use the proportionate change that will get rid of any fraction or turn the last digit into a 0—when possible, of course.

When dealing with the number 45 as a divider or multiplier, we double it—to form the easier-to-handle 90.

When dealing with 33 ⅓, we triple—because it gets rid of the fraction, but also because in this case it turns the number into 100.

How would this rule apply to the divider 3½?

In order to get rid of the fraction, you double it to 7. If you see it in decimal terms—3.5—the .5 is also the signal to double, and you still change it to 7.

How about the number 1.25 (or 12.5, 125, etc.)? If you double it, you have 2.5. This is simpler to handle than 1.25, of course, but it is still in two digits. Double it again, or quadruple the original number, and it becomes the easy-to-handle number 5. You then compensate, if you are dividing, by multiplying the number divided or the answer by 5 to keep the change proportionate.

Changing Downward

Proportionate change does not always mean multiplying. It can also mean changing in the opposite direction. Consider this problem:

$$18\,\overline{)720}$$

How can you most easily change this divider into a single-digit number? You could do it by multiplying by 5, which converts 18 to 90. In this case, however, that is the hard way. Instead, cut it in half. Half of 18 is 9. Keep the change proportionate by cutting the number divided or the answer in half too. $9\,\overline{)360}$ gives 40. Or $9\,\overline{)720} \times \frac{1}{2}$ gives 40.

Here is another example:

$$21\,\overline{)168}$$

No multiplication of the divider will change it to a single-digit number. But cutting it to ⅓ will; ⅓ of 21 is 7. So divide 7 into ⅓ of 168 (56) or into 168 and compensate by dividing the answer by 3.

It may have struck you that dividing a number to convert it is really only a new facet of the factor short cut. It is indeed. When we cut a divider to ½ or ⅓ or some other fraction, we are really factoring it—but you will note that the rest of our handling is a little different, and your frame of mind as you look at the problem is quite different. You are thinking of change—not factors.

Some of the following numbers can be simplified by changing upward, some by changing downward. Play with them a bit until you feel you have the simplest form of each.

15	75	2 ½	2 ⅓	16
3 ⅓	3 ½	12	4 ½	1 ⅓

Try them yourself before reading on.

The quickest way to convert each of these in the proportionate-change short cut is:

15:	double it to 30.
75:	quadruple it to 300.
2 ½:	double it to 5.
2 ⅓:	triple it to 7.
16:	cut it in half to 8.
3 ⅓:	triple it to ten.
3 ½:	double it to 7.
12:	cut it in half to 6.
4 ½:	double it to 9.
1 ⅓:	triple it to 4.

Multiplying

It has already been pointed out that proportionate change applies as easily to multiplying as it does to dividing. In multiplying, however, you reverse the compensation. If you double the multiplier to simplify it, then you cut the other number or the answer in half. If you use ⅓ of the multiplier, then you triple the other number or the answer.

Here is an example:

$$\begin{array}{r} 328 \\ \times 15 \\ \hline \end{array} \quad \text{becomes} \quad \begin{array}{r} 164 \\ \times 30 \\ \hline \end{array} \quad \textit{or} \quad \begin{array}{r} 328 \\ \times 30 \\ \hline \end{array}$$

(divide answer by 2)

Try this one on your pad or in your head:

```
 6 9 5
×4 5
```

Cover the answer with your pad until you have finished.

To simplify 45, you naturally double it to 90. Note here that if you divide 695 by 2, you will have a remainder. This will be wiped out when you multiply by 90; if it is not, then you went astray somewhere. In such a case, however, it is usually easier to divide the answer by 2 rather than the number multiplied.

Here are both workings:

```
Usual way        Two proportionate change ways

 6 9 5           6 9 5                3 4 7.5
×4 5            ×9 0                   ×9 0
---------       ---------           -----------
2 7 8 0         5 2 5 5 0           2 1 2 7 5.0
  3 4 7 5       3 1 2 7 5 (½
---------                answer)
2 0 2 7 5
```

When you change your multiplier by cutting it in half or into another fraction, then you compensate by multiplying the other number or the answer by the same amount. Again, it is just the reverse of your compensation in division.

```
 4 5 3               9 0 6           4 5 3
×1 6    becomes       ×8      or       ×8
-----               -----           -----
                             (double answer)
```

In order to master the point thoroughly, it would not hurt to work out all three forms of this problem. It will help you "feel" the identity of the end results, no matter how the numbers were twisted and turned in working out those results.

Proportionate change is especially valuable in dealing with fractions of all kinds. Even when a proportionate change cannot reduce one of the numbers you must handle to a single digit, it can often simplify it to a remarkable degree.

Would you rather divide by 4 ⅓—or by 13, and multiply the answer by 3?

Is it easier to multiply by 6 ⅝—or by 53, and divide the answer by 8? For an even more dramatic example, you would prefer to handle 6 ¼ as 25—and compensate by multiplying by 4 (in division) or dividing by 4 (in multiplication).

Try out the idea on these numbers:

2 ⅜ 6 ¼ 3 ½ 4 ½ 5 ½

If you multiply each of these numbers by the quantity that will convert it into a whole number, you get the following results (The multiplier is in parentheses.):

2 ⅜ becomes 19. (8)
6 ¼ becomes 25. (4)
3 ½ becomes 7. (2)
4 ½ becomes 9. (2)
5 ½ becomes 11. (2)

In each of the above cases, of course, you compensate in the other number or in the answer with the same multiplier. Do whichever comes more easily.

Run through one whole problem now. Cover the answer with your pad as you work this out with proportionate change:

$$7\tfrac{1}{2}\,\overline{)\,3\ 0\ 0}$$

Surely it is much simpler with the short cut than without it. We double 7 ½ to make 15, which we see at once goes into 300 exactly 20 times—times 2 is 40. Or 15 goes into 600 precisely 40 times.

Now do a multiplication with this technique. Move your pad over the answer and solve this problem with proportionate change:

$$\begin{array}{r} 2\ 8\ 4 \\ \times\ 1\ \tfrac{3}{4} \\ \hline \end{array}$$

Use whichever proportionate change suits you best, but do it before checking with the answer.

The logical conversion for 1 ¾ is to multiply it by 4 and divide the other number or the answer by the same factor. Your answer either way is 497. The two workings are these:

$$\begin{array}{r} 2\ 8\ 4 \\ \times\ 7 \\ \hline 1\ 9\ 8\ 8 \end{array} \qquad \textit{or} \qquad \begin{array}{r} 7\ 1 \\ \times\ 7 \\ \hline 4\ 9\ 7 \end{array}$$

4 9 7 (¼ answer)

Proportionate change is very largely a special application of factoring, and contains some elements of aliquots as well—as you have no doubt observed. It is such a special application, particularly in compensating in the other number and in often making a number larger, that it is classically considered a separate short-cut method.

As an exercise in number sense, consider the essential identity of doubling 35 to make 70 and factoring it into 7 and 5. In multiplication, if you double 35 to 70, you divide the other number or the answer by 2 in order to compensate. Now, dividing by 2 is an aliquot approach to multiplying by 5, is it not? And we picked up an extra 0 when we doubled 35 to 70—which corresponds to the seemingly missing 0 if we consider a division by 2 to be an aliquot for 5.

The various short cuts overlap and are overlapped by the others in many respects. The basic number relationships remain constant; we are merely using different conversions to make those relationships more visible and easier to handle.

17

CHOOSING AND COMBINING SHORT CUTS

YOU have learned and practiced the four most generally useful short cuts. There are others, but they are quite specialized. The most complete assortment can be found in the books listed in the bibliography. With the four short cuts you have learned, however, you can convert a great deal of your multiplying and dividing into simpler forms.

Review all together in one place the four different approaches:

BREAKDOWN For one of the numbers to be multiplied, use a round number if this permits an adjustment with an easy fraction of the other number *or* of the result of the first step. 39 becomes 40 − 1; 45 becomes 50 − $\frac{1}{10}$ the first product.

ALIQUOTS When one of the numbers is an even fraction of a ten-base, use the fraction instead of the number. 25 is treated as $\frac{1}{4}$ of 100.

FACTORS When one of the numbers can be factored, multiply or divide by each of the factors in turn. 63 is treated as 9, then 7.

PROPORTIONATE CHANGE When one of the numbers can be simplified by doubling or halving it (or any other such change), use the simpler form and compensate the other number or the answer. 35 becomes 70, with a compensating factor of 2.

Many numbers can be short-cut with not just one, but with two or more of these methods. 45, for instance, can be factored (5 × 9), broken down (50 less 1/10 product), or changed (90, compensate with 2). An interesting exercise is to locate one number to which all of these methods can apply. One such number is 125. Witness the various short-cut handlings of the number 125:

BREAKDOWN: 100, plus 1/4 product.

ALIQUOTS: 1/8 of 1,000.

FACTORS: 5 × 5 × 5.

PROPORTIONATE CHANGE: quadruple to 500.

For real mastery of short cuts, try to get a feel for the real *identity* of these four apparantly different relationships. One of the techniques will work out in even-number terms for a number that none of the others might handle in this way, but essentially they are all merely different expressions of the same fundamental situation.

The aliquot approach to 125, for instance, is to take 1/8 of 1,000. The proportionate change approach is to use 500, and compensate by a factor of 4. 500 is just half of 1,000, and 4 is just half of 8. The relationships are the same; only the facets we choose to see in any one case appear to be different.

Numbers to which all four approaches apply without remainders or fractions are few, but numbers for which two of the short cuts work are plentiful. Exercise your understanding of the four approaches by converting each of the following numbers in at least *two* ways:

75 72 36 4½ 63 384

Cover the explanations with your pad until you have found two or more short cuts for each of these numbers.

Here are the possibilities:

75: $3 \times 5 \times 5$; $\frac{3}{4}$ of 100; $300 \times \frac{1}{4}$
72: 8×9; $80 - \frac{1}{10}$ product
36: 6×6; 4×9; $40 - \frac{1}{10}$ product; $30 + \frac{2}{10}$ product
$4\frac{1}{2}$: $5 - \frac{1}{10}$ product; $9 \times \frac{1}{2}$
63: 7×9; $70 - \frac{1}{10}$ product; $60 + \frac{1}{20}$ product
384: $6 \times 8 \times 8$; $400 - \frac{4}{100}$ product

This does not mean that the different short cuts are of equal value whenever a number can be converted in two or more different ways. The value in each case depends not only on the possibilities of that number, but also on the other number involved. It also depends on which of the ways you find most adapted to your own ease and speed. The idea is to pick the simplest method for working the problem. Sometimes it will be one of the short cuts, and other times it will be to go ahead and do it with your new streamlined arithmetic. Flexibility is the key. You do not dig a hole for a rose bush with a steam shovel, or use a garden spade for a house foundation. Equally, you do not use three two-digit factors in place of multiplying by the number you factored, because it would not save you any work.

Combining Short Cuts

The possibilities of combining two short cuts in one problem are quite extensive and rather intriguing.

Breakdown by itself, for instance, makes sense only if you can break a number down to a simple base and an adjustment that is a simple fraction of the other number or the product. If you combine methods, however, you can use any aliquot or any easily factored or any easily changed number as a base.

In the range of numbers in which breakdown alone would save you work, 25 was not a useful base because you would still have to multiply through by two digits. But with aliquots to help, you could break down 26 into the combination technique "divide by 4 (aliquot) plus the other number (breakdown)."

Here is how you would do it:

Usual way	Aliquot-breakdown way
8 9 3	2 2 3 2 5 (¼ of 89300)
× 2 6	+ 8 9 3
1 7 8 6	2 2 1 1 8
4 3 5 8	
1 2 1 1 8	

Note that since you are dividing by the easy-to-handle divider 4, your short-cut method is to jot down only the answer.

Surprisingly difficult-looking problems can sometimes be solved almost at sight when one of the numbers happens to contain an aliquot as an easy breakdown base. 1375 × 8642 becomes merely ⅛ of 86420000, plus $\frac{1}{10}$ of the product—because 1375 can be broken down into 1250 (⅛ of 10,000) plus 125 ($\frac{1}{10}$ of 1250). Try it and see.

Do the following problem with an aliquot-breakdown.

```
  4 7 8 2
× 3 8 5
```

Cover the explanation with your pad while you give it your best.

We break the number 385 into 375 plus 10—⅜ of 1,000 plus 10 times the other number. First you divide 4782000 by 8, jotting down only the answer. Multiply the result by 3. Then add 47820—10 × 4782:

```
5 9 7 7 5 0
        × 3
1 7 9 3 2 5 0
  + 4 7 8 2 0
1 7 3 0 0 7 0
```

Just for comparison, here is the usual way of solving the same problem:

$$\begin{array}{ccccccc} 4 & 7 & 8 & 2 & & & \\ \times & 3 & 8 & 5 & & & \\ \hline 1 & 4 & 3 & 4 & 6 & & \\ & 3 & 7 & 2 & 5 & 6 & \\ & & \bar{2} & 3 & 9 & 1 & 0 \\ \hline 1 & \underline{7} & 3 & \underline{\underline{9}} & 0 & 7 & 0 \end{array}$$

Try breaking down these multipliers to aliquot bases:

376	124	26	865
275	55	2.25	1375

Some of these get a little tricky, but each of them can be broken down to an aliquot base. Here is how:

376: $\frac{3}{8}$ of 1,000, plus the other number
124: $\frac{1}{8}$ of 1,000, minus the other number
26: $\frac{1}{4}$ of 100, plus the other number
865: $\frac{7}{8}$ of 1,000, minus 10 times the other number
275: $\frac{1}{4}$ of 1,000, plus $\frac{1}{10}$ the product
55: $\frac{1}{2}$ of 100, plus $\frac{1}{10}$ the product
2.25: $\frac{1}{4}$ of 10, minus $\frac{1}{10}$ the product
1375: $\frac{1}{8}$ of 10,000 plus $\frac{1}{10}$ the product

Just as you can break down a complex multiplier to an aliquot base as well as a rounded-off base, so can you break down multipliers to a factorable base. There would be no point in breaking down 37 by the breakdown method alone. But by combining it with the factor short cut, 37 becomes 6 × 6 (factors) plus the other number (breakdown).

See if you can find the surprisingly easy breakdown-factor short cut for this problem:

$$\begin{array}{r} 4693 \\ \times\ 8181 \\ \hline \end{array}$$

Any series of digits that repeats itself—such as 81, 81—is a very automatic breakdown. This number is 8100 plus $\frac{1}{100}$ of itself. 8100, in turn, is at sight 90 × 90. So the breakdown-factor short cut would be: "90 × 90 (factors) plus $\frac{1}{100}$ of the product (breakdown)." In addition, however, we often handle multiplication by 9 as a breakdown, using 10 − 1. In this case, 90 is 100 − 10. Let us show all three methods of working:

Usual way

```
   4 6 9 3
 × 8 1 8 1
 ---------
   3 6 5 4 4
     0 4 6 9 3
       3 6 5 4 4
         0 4 6 9 3
   -----------------
   3 7 2 7 2 3 3 3
```

Breakdown-factor ways

```
4 6 9 3                       4 6 9 3 0 0
  × 9 0               Or    − 4 6 9 3 0
-----------                 -----------------
3 1 2 3 7 0                 4 2 3 4 7 0 0 0
        × 9 0               − 4 2 2 3 7 0 0
---------------             -----------------
3 7 9 0 3 3 0 0             4 8 0 1 4 3 0 0
  + 3 8 0 1 3 3                 + 3 8 0 1 3 3
---------------             -----------------
3 7 3 9 3 4 3 3             3 8 3 9 3 4 3 3
```

As very often happens, the illustrations of the two short cuts do not dramatize the real simplification involved: the handling of easier processes at each step. Follow each of them pencil in hand to see how this works.

Breakdown can also sometimes be combined with proportionate change. The number 34 does not find a natural place in any one of the individual short-cut methods. But if you realize that 34 is just one less than an easy proportionate-change base, you might choose to handle it as a multiplier as 70 (proportionate change) and cut the other number or answer in half; then subtract the other number (breakdown).

You might or might not choose to use any of these specific combinations. Again, and again: hunt for relationships, use the short cut or combination of short cuts that flashes into your mind as an easy and sensible method, and get the problem done. This, after all, is the end purpose of all mathematics, short-cut or not; get the problem done.

Other Combinations

The possible combinations of short cuts are almost endless. A book could be written about the refinements of double and triple and quadruple combinations of methods. It would be an interesting exercise, but would not really get you through your arithmetic with greater speed and accuracy except for the particular relationships that happen to hit you with special and memorable force.

One or two other wrinkles would, however, speed up your number work from time to time. They are rather intriguing, too.

We noted a page or two back that sometimes you will break a multiplier down to a factorable base. You will, as well, discover sometimes after you have factored a number that one of the factors is too complex to save much time in using the straight factor approach—but that complex factor might be broken down. This is just the *reverse* of breakdown-factor; it is, if you will, factor-breakdown.

Let us try one. As a start, factor the multiplier 261.

The digit sum of 261 is 0, so you know 9 is a factor. 9 into 261 gives you 29 as the other factor.

Now 29 is a prime number and it is a two-digit number, so factors do not short-cut this problem as much as we should like. *But* 29 is a very natural candidate for breakdown. We might solve a multiplication involving 261 by multiplying by 9, then 30, then subtracting.

Be very careful here to subtract, not the other number, but the product of 9 times the other number. Why? Because that $30 - 1$ is a factor, not a breakdown of the whole number. Work the factors backward, if you wish, to get this point clear. 9×30 is 270. Subtract 9 (not 1) from 270 to get 261, the number we started with.

Here is an example involving this specific factor-breakdown:

Usual way	Factor-breakdown way
5 8 7	5 8 7
× 2 6 1	× 9
1 7 7 4	4 2 8 3
3 4 2 2	1 5 8 4 9 0 (above product tripled, plus 0)
0 5 8 7	− 5 2 8 3
1 4 2 1 0 7	1 5 3 2 1 7

We have covered only the combinations involving breakdown because they are the most generally useful. Some combinations do not make any sense at all, such as factoring to an aliquot. Play with the idea on your pad and you will see why.

The ultimate short cut is to have so firm a grip on your number sense and on the possible short cuts—together with useful combinations of them—that in each case you can quickly and unerringly pick the shortest, easiest road to the solution.

The next step is obvious. It is to practice, on some actual examples, the best approach to each. Do not bother to solve the following problems unless you wish to. The exercise is simply to select, in each case, the best technique. Keep in mind as you go through the exercise that in multiplication you might choose to short-cut either number, not just the one that appears on the bottom.

Examine each of the following problems for all reasonable short-cut possibilities and definitely state to yourself how you would tackle it before going on to the suggested approaches. Not all of them, by the way, should be converted. In four cases, there is no short cut possible. For practice, however, spend more time with each than you would expect to spend looking for short cuts in your work with figures.

1 1250×67

2 271×137

3 491×9

4 $35\,\overline{)\,4896}$

5 983×126

6 $\frac{4}{5} \times 436$

7 632×47

8 357×79

9 $\frac{2}{5} \times 689$

10 536×99

11 $378\,\overline{)\,64937}$

12 607×69

13 $72\,\overline{)\,8935}$

14 683×45

15 1324×75

16 180×97

17 984×625

18 $25\,\overline{)\,7654}$

19 $375\,\overline{)\,48296}$

20 841×432

21 245×241

22 317×367

23 $49\,\overline{)\,9364}$

24 3.9×4.5

25 $223\,\overline{)\,6843}$

26 256×239

27 443×307

28 $225\,\overline{)\,7398}$

29 $875\,\overline{)\,34672}$

30 317×63

Don't skip over the above exercise. Short of knowing the short cuts themselves, it is the most important practice in the short-cut section of this book. It does little good to know several short methods if you cannot see quickly whether or not each can be used.

In some of the above problems more than one conversion can be applied. You can treat the divider 25 in problem 18, for instance, as 5 × 5, or ½ of 50, or ¼ of 100. The suggested short cuts below, however, are those I believe simplest in each case. You are perfectly free to choose a different one if it will work and if it is easier for you.

1. Convert the top number into ⅛ of 10,000.
2. No practical short cuts. Do it straight.
3. All short cuts are not complicated. It is still easier to multiply by 9 by subtracting the number from 10 times the number.
4. You can factor 35 into 7 and 5, or double it to 70.
5. Two-step short cut. 126 is 125 plus 1, and 125 is ⅛ of 1,000.
6. Here is a reverse aliquot. Far simpler to multiply by .8 than by ⅘.
7. 47 is 1 less than 48, which is 6 × 8.
8. Treat 79 as 80 − 1.
9. Reverse this fraction to its aliquot form: .4.
10. The most elementary of all short cuts. 99 is 100 − 1.
11. Factor 378 into 9, 7, and 6. You divide three times, but by a single digit each time.
12. 69 is, of course, 70 − 1.
13. Factor the 72 into 9 and 8.
14. Choose among factoring the 45 into 9 and 5; doubling it to 90; or breaking it down to 50 − ⅒ product.
15. 75 is ¾ of 100. Instead of multiplying by 75, just multiply by 3(00) and divide by 4.
16. 1 would convert the top number on this, although 97 is easy as 100 − 3. But 180 is twice 90, so subtract 10 97's from 100 97's and double the answer.
17. 625 is an aliquot, being ⅝ of 1,000.
18. Don't ever divide by 25. Subtract two zeros from the number divided (using a decimal) and multiply by 4.
19. 375 is ⅜ of 1,000. Subtract three zeros; multiply by 8; then divide by 3.

20. The digit sum of 432 is 0, so you can factor it: 9, 8, and 6.
21. You cannot do much to the bottom number, but 5 is obviously a factor of the top number. A quick sight-division shows that the other factor is 49, which in turn you factor to 7 and 7.
22. No practical short cut.
23. You should recognize factorable numbers of 81 or less at a glance. 49 is 7 × 7.
24. Do not let the decimal fool you. 4.5 can be handled just like the 45 in problem 14—but keep track of the decimal point.
25. No short cut would be worthwhile here.
26. Factor the top number in this problem. 256 is the product of 8, 8, and 4.
27. This is the last of the booby traps. Use your no-carry, left-to-right multiplication for quick results.
28. This divider can be converted to a single digit with proportionate change. Divide by 900 and multiply by 4.
29. 875 is a perfectly good aliquot. Subtract 4 zeros, then divide by 8 and multiply by 7.
30. 63 is an easy breakdown: 70 − $\frac{1}{10}$ product.

18

MASTERING FRACTIONS

OF ALL the specialized branches of mathematics, fractions seem to be greeted with more general panic than all the others put together.

It does not have to be so. Fractions are really not much more complicated than multiplying or dividing. Perhaps the reason for their general unpopularity is that they are taught, to an even greater extent than is true for the other processes, almost entirely by rote. The rote itself simply has to have a few more steps and rules than do whole numbers.

You can add any two whole numbers together without doing anything to them first. But not fractions. The reason *why* this is so has apparently escaped the normal teaching methods. Many people have trouble understanding why you can multiply two fractions together and get an answer smaller than either of them. If you multiply two numbers together, isn't the answer larger than either? Again—not with fractions.

Both peculiarities, along with the other peculiarities, are inherent in the true nature of fractions. Let us approach their nature with some general observations.

A fraction is, in essence, a number that cannot be expressed normally in our decimal system of digits running from 1 to 9 and then starting over. It is usually *smaller* than 1, and our counting system has no way of expressing such a quantity other than the apparently awkward form of the fraction (other than a decimal, which is merely a fraction written in another way).

A fraction, even if we have no other way to indicate it than a fraction, is however a very real number or quantity. The form in which we show it is really a fabulously ingenious and useful method of expressing any conceivable quantity *from any conceivable counting base* in terms of the number system we know.

Imagine, if you will, that our base quantity "1" is a loaf of bread. We have built up a complete arithmetic based on loaves of bread; we have units of ten loaves; we have learned by heart how to add 3 loaves to 6 loaves, to start with 8 loaves and take away 4 loaves, to imagine that one group of 2 loaves has been doubled, or multiplied by 2. But then, suddenly, one of our loaves breaks into pieces and we must account for the pieces.

This is a fraction. The loaf may have broken into "3" pieces, but we have no arithmetic with which to handle it. The only units we know are in terms of loaves of bread. Yet this "1"—this loaf—is no longer 1. It is less than 1.

How do we express the quantity represented by each of these pieces? Some genius or geniuses, centuries ago, suggested that we represent it by "1"—because it once was 1 loaf—*divided by* "3"—as if each of the pieces were now a loaf. The 3 came from 1, so the essential quantity is the one expressed by a division of 3 into 1.

We write it $\frac{1}{3}$.

This is the basic fact about all fractions. They are real quantities, but quantities that cannot be expressed in our regular number system, so we express them in terms of divisions.

A fraction is, then, merely a division problem.

When we write the quantity $\frac{2}{5}$, we really intend to convey the idea of a quantity that is outside our number system, and can best be expressed by dividing 2 by 5, or $2 \div 5$, or $5\overline{)2}$. Because we wish to show it as a quantity more than a problem, we write it $\frac{2}{5}$.

Thinking of a fraction as really a *problem in division,*

which also expresses a specific quantity, may help you to gain an emotional grasp of the entire system.

Why 2 × 2 *Is "Less" Than* 2

One of the most baffling habits of fractions is that when you multiply two of them together, your answer is less than either of them alone. We are so accustomed to thinking of multiplication as an increasing process that this jars our basic number sense.

If you think of multiplying as counting a number a certain number of times—which is precisely what it is—the concept becomes clearer. If you count a number more than once, then the result is obviously larger than the number was. But if you count the number *less* than once, as you do when you count it only ⅓ times, for instance, then the answer must be smaller than the number was when you started. If the number you counted was less than 1 to start with, such as ¼, then the answer will obviously also be smaller than the number of times you counted it—because to get an answer as large as your "counting" number you would have to count another number at least as large as 1.

This is why multiplying by two fractions smaller than 1 gives you an answer smaller than either of the fractions. You count a number that is smaller than 1 to begin with, and you don't even count it one whole time. When you multiply ¼ × ½, you are saying in effect "count ½ exactly ¼ times."

This fact leads us into the first natural rule for handling fractions with understanding as well as memorized rules: to multiply fractions, multiply the top numbers together for the top of the answer, and multiply the bottom numbers together for the bottom of the answer.

Note that we define this rule in terms of top numbers and bottom numbers. Arithmetic has become topheavy with special names such as "numerator" and "denominator" that confuse things more than they clarify them for most of us. If you agree, "top" and "bottom" is instantly and unmistakably clear.

Following this rule, then, count ⅗ exactly ¼ times—or, if it sounds clearer, ¼ of one time:

$$\frac{3}{5} \times \frac{1}{4} = \frac{3}{20}$$

The top of our answer is 3, which is what we get when we multiply 3×1. The bottom is 20, which is produced by multiplying 5×4. This is what is produced when you start with a quantity expressed by dividing 5 into 3 ($3/5$) and count it not even once, but a number of times expressed by the division of 4 into 1 ($1/4$).

In order to refresh your memory, try it yourself:

$$\frac{3}{4} \times \frac{2}{7}$$

This is simple, naturally, but if you are at all rusty it would help to cover up the answer with your pad and write down the answer.

Multiplying the top numbers, we get 6. Multiplying the bottom numbers, we get 28. The answer is $6/28$.

This is true, but $6/28$ is a fairly complex fraction. Is there a simpler expression of the same quantity? 6 and 28 are both evenly divisible by 2. If we divide both the top and bottom by 2, our fraction becomes $3/14$.

Think about this fact for a bit. Your memory of the rules undoubtedly tells you that it is the same, but visualize the two expressions and see if you can feel their identity.

This leads us to a general rule for all fractions:

> If you multiply or divide both the top and bottom numbers of a fraction by the same number, the quantity remains unchanged.

By this rule, $6/8$ is the same quantity as $3/4$. Is it? You know by training that it is. But can you *feel* it? As a good exercise in number sense, try expressing this quantity by 6 dots above a line with 8 dots below it. Thoughtfully connect each two adjacent dots so they become 1 line, in pairs, and note that you now have 3 lines over 4 lines. You have not changed the *relationship* of the quantities above and below the line, but you have changed the numbers.

Try a few multiplication exercises on your pad:

$$\frac{3}{8} \times \frac{2}{9} \qquad \frac{1}{6} \times \frac{5}{12} \qquad \frac{1}{2} \times \frac{3}{4} \qquad \frac{1}{4} \times \frac{1}{2}$$

Work these out. They are elementary, but important.

The raw answers are, of course, $6/72$, $5/60$, $3/8$, and $1/8$. We say "raw" because some of these can be reduced to simpler terms. $6/72$, for instance, is reduceable at sight to $3/36$, and this in turn is clearly $1/12$. Check your other answers for reduction possibilities.

Short-Cut Multiplying

If any fraction whose top and bottom numbers can be evenly divided by the same number can be reduced to a simpler form by dividing, then two fractions to be multiplied can also go through the same process even before they are multiplied.

This means that often you can do part of the reducing before you multiply, rather than after.

The secret that makes this possible is that it does not make a bit of difference in what order you multiply or divide numbers: the result will be the same. $4 \times 8 \times 6$ is the same as $8 \times 6 \times 4$ is the same as $6 \times 4 \times 8$—as well as $8 \times 4 \times 6$ and $6 \times 8 \times 4$.

If this fact is not instinctive with you, work out each of the above multiplications and make it instinctive.

When we start out with a problem such as the first one above, we note that more than one top and bottom can be divided by the same number:

$$\frac{3}{8} \times \frac{2}{9}$$

The top 3 and bottom 9 are both divisible by 3. They become, respectively, 1 and 3. So now we have:

$$\frac{1}{8} \times \frac{2}{3}$$

But the top 2 and bottom 8 are also divisible by the same number. Dividing both by 2, we have:

$$\frac{1}{4} \times \frac{1}{3}$$

Now our answer comes out as $\frac{1}{12}$. This is the same as our answer the first time we tried it, after reduction.

This process is traditionally called "cancellation." It might more sensibly be called "reduction," because that is what you really do. You do not cancel anything; you reduce numbers where you can.

Try the short-cut method of reduction on these:

$$\frac{4}{7} \times \frac{5}{12} \qquad \frac{3}{5} \times \frac{5}{9} \qquad \frac{2}{3} \times \frac{6}{7} \qquad \frac{5}{6} \times \frac{3}{20}$$

Use your pad to play with these before checking the reduced forms below.

$\frac{4}{7} \times \frac{5}{12}$ can be reduced by dividing the top 4 and bottom 12 by 4. The reduced form is $\frac{1}{7} \times \frac{5}{3}$, giving the answer $\frac{5}{21}$.

$\frac{3}{5} \times \frac{5}{9}$ has two reductions. The two 5's can both be divided by 5, which gives us $\frac{3}{1} \times \frac{1}{9}$. The top 3 and bottom 9 can both be divided by 3, giving $\frac{1}{1} \times \frac{1}{3}$. The answer must be $\frac{1}{3}$.

$\frac{2}{3} \times \frac{6}{7}$ offers only the bottom 3 and top 6, both to be divided by 3. Now the problem is $\frac{2}{1} \times \frac{2}{7}$, which gives $\frac{4}{7}$.

$\frac{5}{6} \times \frac{3}{20}$ offers two reductions. 5 goes into the top 5 and bottom 20, reducing the problem to $\frac{1}{6} \times \frac{3}{4}$. 3 goes evenly into the bottom 6 and top 3, further reducing the problem to $\frac{1}{2} \times \frac{1}{4}$. Answer, $\frac{1}{8}$.

There is an important reason why I refuse to call this process "cancellation." The technique is usually taught as an "X" process, from the top of one fraction to the bottom of another. It is definitely not necessarily so; *any* top and bottom (never a top and top or bottom and bottom, of course) will do, in the same fraction or in any of the fractions to be multiplied.

You can reduce $\frac{6}{8} \times \frac{2}{12}$ the same way you would $\frac{6}{12} \times \frac{2}{8}$, using any top and bottom that can be divided evenly by the same number. The only difference is that usually fractions are presented in arithmetic books already reduced for such problems. In our real-life figure work, they are not always so reduced for us. Look for reducing possibilities everywhere.

Dividing Fractions

Just as it may seem peculiar to multiply two quantities (if they are fractions) and get an answer smaller than either of them, so may it appear outrageous to divide one quantity into another and (if they are fractions) get an answer larger than either.

Keep firmly in mind that division is merely the reverse of multiplication, and review in your mind the reasons for the strange results you get in multiplication. In effect, the fraction divided is the answer to an imaginary multiplication, and the purpose of the division is to find the missing partner in the multiplication.

Let us start into the division of fractions with a simple example:

$$\frac{3}{4} \div \frac{1}{2}$$

If we multiply by multiplying the respective tops and bottoms, then we might expect to divide by dividing them. In a problem this simple, we can indeed: 1 into 3 gives 3, and 2 into 4 gives 2: $3/2$ is the answer.

Do not worry about that $3/2$ yet. We will get into so-called improper fractions later.

The technique of simple division will theoretically work with any problem, but since every number does not "go into" every other number evenly we sometimes would end up with awkward decimal remainders and create some really difficult-to-handle answers.

This is why the standard trick of "inversion" has been developed. The trick has this rule:

> To divide by a fraction, turn it upside down and multiply by it.

If this seems at all odd, reinforce your grasp of the reason why, as well as the rule, by considering that all division is merely an inversion of multiplication. When you multiply by 4, you count the other number 4 times. When you divide by 4, you count the other number $1/4$ times.

Another way of saying "divide by 27" is to say "multiply by $\frac{1}{27}$."

So another way to say "divide by $\frac{3}{4}$" is to invert the fraction and say "multiply by $\frac{4}{3}$."

The single greatest source of confusion to many people is remembering which fraction to invert. If you fully understand the why, you cannot ever again become confused. To make extra sure, run through the comparison once more.

In order to divide 28 by 14, would you set it up as

$$14 \times \frac{1}{28} \quad \text{or} \quad 28 \times \frac{1}{14}?$$

So in order to divide $\frac{1}{2}$ by $\frac{1}{4}$, would you set it up as

$$\frac{1}{4} \times \frac{2}{1} \quad \text{or} \quad \frac{1}{2} \times \frac{4}{1}?$$

It is the fraction *by* which you divide that you invert —always.

$$\frac{4}{5} \div \frac{1}{3} \quad \text{becomes} \quad \frac{4}{5} \times \frac{3}{1}$$

Pull out your pad and do these examples with inversion:

$$\frac{3}{4} \div \frac{1}{2} \qquad \frac{7}{8} \div \frac{2}{5} \qquad \frac{5}{6} \div \frac{2}{3}$$

Inverting the divider of the first problem gives us $\frac{3}{4} \times \frac{2}{1}$. The answer is $\frac{6}{4}$, which reduces to $\frac{3}{2}$. If you used short-cut reduction, you would have converted the problem to $\frac{3}{2} \times \frac{1}{1}$ before multiplying.

The second problem becomes $\frac{7}{8} \times \frac{5}{2}$, which gives an answer of $\frac{35}{16}$. This fraction cannot be reduced.

The third, when you invert the divider, becomes $\frac{5}{6} \times \frac{3}{2}$. This can be reduced to $\frac{5}{2} \times \frac{1}{2}$, with the answer $\frac{5}{4}$.

Short-Cut Division

If you are sure of your technique, there is no need to rewrite such division problems with the divider inverted. You can do the inversion in your head by following this rule:

> To divide, multiply the top of the fraction divided by the bottom of the divider, and put it on top. Multiply the bottom of the fraction divided by the top of the divider, and put it on the bottom.

In other words, you simply multiply each top by the other bottom. Keep your answer straight by using the fraction divided as your guide for the answer: the product of this top and the other bottom becomes the top of the answer. The entire process automatically inverts the divider without rewriting.

Here is an example:

$$\frac{2}{3} \div \frac{3}{4}$$

Top of fraction divided (2) times the other bottom (4) is 8. Since you used the top of the fraction divided, this 8 goes on top of the answer. Bottom of fraction divided (3) times the other top (3) is 9. This goes on the bottom of the answer. The answer is $8/9$.

Try one yourself:

$$\frac{1}{4} \div \frac{2}{5}$$

Keep your top and bottom straight by matching to the fraction divided rather than to the divider, and you can read the answer at sight. It is $5/8$.

Beware of one possible misunderstanding here. When you divide in this fashion, you *cannot reduce* in the normal fashion by dividing tops and bottoms simultaneously. This is because you would invert the divider if you rewrote it before multiplying, so in essence the top of the divider becomes its bottom and vice versa. You can, if you take care to keep track of the proper tops and bottoms, reduce by dividing both tops or both bottoms by any number that will go into them evenly, because you invert the divider in multiplying anyway.

Adding Fractions

Adding and subtracting fractions is, surprisingly, more

work than multiplying or dividing them. The reason is simple, and is based on the fact that it makes no difference in what order you multiply a series of numbers—but it makes a big difference in what order you multiply and add.

Consider this quantity:

$$2 \times 3 + 4$$

Does it make any difference whether you treat this as 2×3, plus 4—or as $2 \times$ the sum of $3 + 4$? Try it and see. One handling gives you 10. The other gives you 14.

It is critically important to add and multiply in the proper portions. $2 \times (3 + 4)$ is not the same as $(2 \times 3) + 4$. Examine the two expressions carefully and you will discover the cause for the difference. In the first handling, the 4 gets multiplied by the 2 after it has been added to the 3. In the second, the 4 never gets multiplied by the 2 at all. So the end result is quite different.

Another way of approaching the special rules for adding and subtracting fractions is to remember that each fraction is, depending on its bottom number, in its own special number system—one not accounted for in our regular digits and expressible in our digit system only as a division problem. $\frac{1}{3}$, $\frac{2}{3}$, and $\frac{3}{3}$ are all quantities based on one-third of 1. But $\frac{1}{4}$, $\frac{2}{4}$, $\frac{3}{4}$, and $\frac{4}{4}$ are quantities based on one-fourth of 1. Thirds and fourths are not in the same number system at all, and trying to add or subtract combinations of the two is like adding gallons and litres.

The first job in adding or subtracting fractions, then, is to get them all into the same number system, Fortunately, it is not hard at all.

There is a very simple way of converting different fractions into the same system. We just multiply the bottoms and adjust the tops. We can even forget that forbidding schoolroom phrase "lowest common denominator," because we do not need it. All we have to do is multiply.

Here is how it works:

$$\frac{3}{4} + \frac{2}{3}$$

First, in order to determine the number system in which we can express both quantities, we multiply the bottoms. 4×3 is 12. This 12 will be the bottom of the answer, because it is a number system that can express both fourths and thirds accurately.

Before we can add, however, we must convert each fraction to this new system. $3/4$ is $3/4$, but it is not $3/12$. How many twelfths is it? The simplest way to convert is to multiply each top by the *other* bottom, because this is the number by which we multiplied the bottom and, as we know, multiplying top and bottom by the same number does not change the value of the fraction.

3×3 is 9, so $3/4$ is $9/12$. We do not worry about that in working, however. All we care about at the moment is the 9. For the second fraction, we multiply 2×4 and get 8. Now we add the two products, and this becomes the top of the answer. $9 + 8$ is 17. The answer is $17/12$.

Once again, look at these four expressions and try to feel their identity:

$$\frac{3}{4} + \frac{2}{3} = \frac{(3 \times 3) + (2 \times 4)}{4 \times 3} = \frac{9 + 8}{12} = \frac{17}{12}$$

This answer is a fraction larger than 1. We will get to the handling of such fractions soon. First, let us finish addition and subtraction.

Try the simplified rule on the following addition. The rule, in one sentence, reads: To add fractions, multiply the bottoms for the new bottom; multiply each top by the other bottom and add these products for the new top.

$$\frac{1}{2} + \frac{2}{3}$$

You do not have to go through the entire step-by-step visualization above each time you do it. Just multiply the bottoms for the bottom of the answer. Multiply each top by the other bottom and add the products for the top of the answer.

For the problem above, our bottom is 2×3 or 6. 1×3 is 3, plus 2×2 is 4, gives 7 as the top. The answer is $7/6$.

Try a few more with this technique. It is really simpler and faster than worrying about common denominators:

$$\frac{1}{4} + \frac{3}{5} \qquad \frac{5}{6} + \frac{1}{2} \qquad \frac{1}{3} + \frac{1}{7} \qquad \frac{2}{5} + \frac{3}{4}$$

Work out and reduce where possible the answers to these on your pad.

The answers, in order, are $^{17}/_{20}$, $^{16}/_{12}$, $^{10}/_{21}$, and $^{23}/_{20}$. The second answer—$^{16}/_{20}$—can be reduced to $^{4}/_{3}$.

Adding More Than Two

The rule becomes just a little more complicated when you add three or more fractions. You have to reduce all of them to the same number system.

The rule is not very much more complicated, however. Let us take it in two steps:

1. Multiply all the bottoms together. This will be the bottom of the answer.

2. Multiply each top by *all* the bottoms *except* its own, and add all the products. This will be the top of the answer.

This rule is precisely the same as the rule for adding two fractions, generalized to handle any number of fractions. Here is an example:

$$\frac{1}{2} + \frac{2}{3} + \frac{3}{5}$$

The first step is to multiply all the bottoms together. 2×3 is 6, $\times 5$ is 30. The bottom of the answer is 30.

The second step is to multiply each top by all the bottoms except its own. 1×3 is 3, $\times$ 5 is 15. 2×2 is 4, $\times$ 5 is 20. 3×2 is 6, $\times$ 3 is 18. Add 15 and 20 and 18 to get the top of the answer: 53. The answer is $^{53}/_{30}$.

Examine carefully the steps in this addition, and you will see that in each case we are really multiplying each fraction's top and bottom by the same number: the products of the bottoms of all the other fractions. This translates all the fractions into the same number system and adjusts all the tops at the same time, without changing the quantity of each fraction.

Do one on your own with this method:

$$\frac{3}{4} + \frac{2}{3} + \frac{2}{5}$$

First, find the bottom of the answer. 4 × 3 is 12, × 5 is 60.

Now for the top. 3 × 3 is 9, × 5 is 45. 2 × 4 is 8, × 5 is 40. 2 × 4 is 8, × 3 is 24. Adding 45, 40, and 24, we get 109 as the top of the answer . . . $^{109}/_{60}$.

Special Cases

There is a further short cut in adding a series of fractions in which some of them are already in the same terms. The usual method is to hunt through all the bottoms for the "lowest common denominator," which takes a bit of figuring and then adjustment of each top.

It is far easier simply to add all like fractions first; then add the resulting unlike fractions in the method just described.

In order to add like fractions (all thirds, say, or all fifths), you simply add the tops. Do nothing to the bottoms. The sum $^1/_5$ and $^2/_5$ is the sum of the tops—3—over the same bottom: $^3/_5$.

Here is how to handle a typical situation:

$$\frac{1}{5} + \frac{2}{7} + \frac{2}{5} + \frac{4}{7}$$

The simplified way is first to add the like fractions. $^1/_5$ and $^2/_5$ are in the same terms, so they total $^3/_5$. $^2/_7$ and $^4/_7$ are in the same terms, so they total $^6/_7$. Now merely add $^3/_5$ and $^6/_7$ as you have done before, and get $^{51}/_{35}$.

Try it yourself:

$$\frac{2}{9} + \frac{1}{5} + \frac{2}{5} + \frac{4}{9}$$

The first and last fractions are both in ninths, so simply add the tops: $6/9$. The second and third fractions are both in terms of fifths, so add the tops and get $3/5$. The sum of $6/9$ and $3/5$ is $57/45$, which reduces to $19/15$.

Reducing As You Go

You can save work by reducing fractions as you go. The $6/9$ in the last example can be reduced at sight to $2/3$. This gives the final answer, $19/15$, directly. It is obviously easier to reduce $6/9$ to $2/3$ than to note that $57/45$ is also divisible, top and bottom, by 3.

It is good practice, then, to reduce each fraction in your problem, or any of the intermediate working figures, to its simplest form before continuing.

Another form of reduction-as-you-go is to avoid multiplying all the bottoms, when you can. Suppose, for instance, you start out to add this problem:

$$\frac{3}{4} + \frac{5}{8}$$

You will get the right answer if you follow the general rule: 4×8 is 32, for the bottom of the answer; 3×8 is 24, and 5×4 is 20, totaling 44 for the top: $44/32$. This reduces to $11/8$.

Note, however, as you look at the bottoms, that 4 goes into 8 exactly twice. If we simply double $3/4$ to $6/8$, the fraction is in the same terms (eighths) as the other. We can then add the tops and find the answer, $11/8$. This is easier.

The intermediate step here, $6/8$, is not the simplest expression of the quantity; $3/4$ is. Yet because it puts the quantity into the same numerical system as the other, $6/8$ is the simplest expression for this problem.

The same lesson applies when you add a series of fractions. The short cut is first to add all like fractions, then add the results. If inspection shows you as you start multiplying the bottoms that your product so far is identical with (or divisible by) one or more of the other bottoms, stop right there and add the fractions so far before continuing.

Here is an example:

$$\frac{1}{2} + \frac{2}{3} + \frac{5}{6}$$

Using the general rule, you start to find the bottom of the answer by multiplying the bottoms. 2 × 3 is 6, times . . . the next number is identical.

This means that the first two fractions can be expressed in sixths. The last fraction is already in sixths. So—this is non-standard, but a definite short cut—first add the two fractions on the left, then add their total to the last fraction. Work it out both ways, if you wish, and see that your final answer is $^{12}/_{6}$ either way. This is a very "improper" fraction indeed, but we will get through subtraction before taking up that subject.

Rather than seeking lowest common denominators, then, start the addition of any series of fractions by multiplying the *smallest* bottoms first. Often, your running product will be identical with larger bottoms when you get to them, or evenly divisible into them or by them. In this case, add up the fractions so far and then add this sum to the others. It is an easier approach.

Your number sense is the best guide to partial addition before completing a problem. If you start to add three fractions with bottoms of 12, 3, and 4, you will note that 3 × 4 is 12. So first adjust and add these two fractions, then add the sum to the other . . . which is already in twelfths.

Subtracting Fractions

If you are completely and confidently at home with adding fractions, subtraction poses no problems. The rules are all identical in reverse. Instead of adding the adjusted tops, you subtract the top of the smaller fraction from the top of the larger fraction. (Larger and smaller applies not to the individual tops or bottoms, of course, but describes which fraction is subtracted from which. It is easier than "minuend and subtrahend.")

One example should make this clear:

$$\frac{3}{4} - \frac{2}{9}$$

Start just the way you would in adding. Multiply the two bottoms to find the bottom of the answer. 4×9 is 36.

Now, however, you clearly separate one top from the other top, because it makes a great deal of difference in subtraction although none in addition. The top of the larger fraction is 3. Multiplying this by the bottom of the other, we have 27. The top of the smaller fraction is 2. 2×4 is 8. 8 from 27 is (complement and slash) 19. The answer is $\frac{19}{36}$.

Do one yourself:

$$\frac{4}{5} - \frac{3}{4}$$

Use your pad to finish this before going on.

The answer, of course, is $\frac{1}{20}$.

Improper Fractions

In many of the examples, we have produced answers such as $\frac{9}{4}$ or $\frac{53}{40}$. If the top of a fraction is larger than its bottom, then the quantity expressed by the fraction is larger than 1. A fraction expressing a quantity larger than 1 is called improper because the quantity is really a whole number plus a fraction.

Nevertheless, we often deal with "improper" fractions, because these are frequently the most convenient ways of expressing the quantities we are handling.

The method for translating an improper fraction into proper form is simple. You merely divide the top by the bottom. A fraction, you recall, is merely a special way of writing a division problem anyway.

In most cases, the answer to the division of an improper fraction will be a number and a remainder. This remainder will be in the same terms as the improper fraction you started with, so it merely becomes the top of the new fraction. The answer to the division becomes the whole number.

Let us try translating the two improper fractions mentioned above into mixed numbers. The two fractions are $\frac{9}{4}$ and $\frac{53}{40}$:

$$4\overline{)9} = 2\tfrac{1}{4} \quad (\text{quotient } 2, \text{ remainder } 1) \qquad 40\overline{)53} = 1\tfrac{13}{40} \quad (\text{quotient } 1, \text{ remainder } 13)$$

Most improper fractions turn out to be 1 plus a fraction when translated to mixed numbers. If you see by inspection that the top of the improper fraction is less that *twice* the bottom, you do not even have to divide to translate it. Merely put down a 1 for the whole number, and subtract the bottom from the top to produce the top of your fractional part. $\frac{17}{12}$, by this short cut, is 1 plus $\frac{5}{12}$—the 5 being produced by subtracting 12 from 17. The reason for this is obvious, since $\frac{12}{12}$ is equal to 1.

Translate these improper fractions to proper mixed numbers:

$\frac{6}{4}$ $\frac{9}{5}$ $\frac{16}{13}$ $\frac{8}{7}$ $\frac{33}{16}$ $\frac{45}{19}$

Cover the answers with your pad, please.

The proper equivalents for the above fractions are

$1\frac{1}{2}$ $1\frac{4}{5}$ $1\frac{3}{13}$ $1\frac{1}{7}$ $2\frac{1}{16}$ $2\frac{7}{19}$

Mixed Numbers

The "proper" form of many quantities equal to more than 1, but not any even whole number, is expressed in the number-plus-fraction form you just created from improper fractions. Often, you must calculate with such mixed numbers.

In adding or subtracting, you simply handle the two parts of each number separately. If the fractional parts give you an improper fraction at the end, translate it. Then add the entire result to the whole part of the answer. Watch:

$$2\frac{5}{8} + 3\frac{7}{8}$$

Handle this as two separate additions. 2 + 3 is 5. $\frac{5}{8}$ + $\frac{7}{8}$ is $\frac{12}{8}$. First, reduce this to $\frac{3}{2}$. Now translate it to $1\frac{1}{2}$. 5 + $1\frac{1}{2}$ is $6\frac{1}{2}$—the final answer.

The principle does not change when unlike fractions are involved:

$$6\frac{2}{3} + 7\frac{5}{7}$$

First we add 6 + 7 to get 13. Our technique for adding $\frac{2}{3} + \frac{5}{7}$ gives is $\frac{29}{21}$, which translates into $1\frac{8}{21}$. Add this to 13 for the final answer, $14\frac{8}{21}$.

Subtraction is handled in the same way:

$$3\frac{3}{4} - 2\frac{2}{5}$$

First, subtract the whole numbers. 3 − 2 is 1. Now use the standard subtraction method on the fractions to get $\frac{7}{20}$. The answer is $1\frac{7}{20}$.

Sometimes, however, we must subtract mixed numbers in which the fraction in the smaller number is larger than the fraction in the larger number. In this case, we make an improper fraction by "borrowing" 1 from the whole part of the larger number. This is just the reverse of translating an improper fraction to a mixed number.

The technique is very simple. After you "borrow" 1, reducing the value of the whole number by 1, you make an improper fraction by merely adding the *bottom and top* of the fraction to make the new top. I have never seen it described like this, but it works like magic. $1\frac{3}{8}$ becomes $\frac{11}{8}$—because 3 + 8 total 11. So $6\frac{3}{8}$ becomes $5\frac{11}{8}$ after you "borrow" 1 from the 6.

Here is a case in which this technique is required:

$$4\frac{3}{16} - 2\frac{7}{16}$$

You cannot subtract the fractions, because you cannot subtract 7 from 3. *Nor* can you use a complement, because the base here is 16, not 10. The solution is to "borrow" 1 from the 4 and translate $\frac{3}{16}$ to $\frac{19}{16}$ by adding top and bottom for the top of the improper fraction. Now the problem looks like this:

$$3\frac{19}{16} - 2\frac{7}{16}$$

The answer is natural now. It is $1\frac{12}{16}$, and the fractional part quickly reduces to $\frac{3}{4}$. Final answer, $1\frac{3}{4}$.

The principle does not change when the fractional part of the problem is in unlike fractions. You still raise the fraction in the larger number by borrowing, if you need to, and then subtract, using the general technique for subtracting:

$$17\frac{1}{3} - 8\frac{7}{8}$$

First borrow 1 from 17 and raise the fraction so you have $16\frac{4}{3}$. 8 from 16 leaves 8. $\frac{7}{8}$ from $\frac{4}{3}$ leaves $\frac{11}{24}$. Answer, $8\frac{11}{24}$.

Multiplying and Dividing

When you come to multiplying and dividing mixed numbers, however, the situation is quite different. This is because multiplying or dividing affects *every* part of every number. If we multiply $14\frac{6}{7} \times \frac{3}{4}$, for instance, we must "count" *both* the 14 and the $\frac{6}{7}$ exactly $\frac{3}{4}$ of one time.

The easiest general rule is to turn every mixed number into an improper fraction when you must multiply or divide. Since you are usually "borrowing" far more than 1 from the whole number—you "borrow" the entire number—you do not just add the top and bottom of the fraction for the new fraction. The rule, however, is not much more complicated:

> To turn any mixed number into an improper fraction, multiply the whole number by the bottom of the fraction, add the top of the fraction, and put the result over the bottom.

Turn $7\frac{3}{8}$ into an improper fraction by this rule. First, multiply the (whole) 7 by the (bottom) 8: 56. Second, add the (top) 3: 59. Put this result over the bottom: $\frac{59}{8}$.

If you try translating $\frac{59}{8}$ back into "proper" form, you will find that it does come out to $7\frac{3}{8}$.

Follow this multiplication:

$$7\frac{2}{3} \times 3\frac{3}{4}$$

Both numbers must first be turned into improper fractions. $7\frac{2}{3}$ becomes $\frac{23}{3}$ (7×3, plus 2). $3\frac{3}{4}$ becomes $\frac{15}{4}$ (3×4, plus 3).

Pause for a moment to see if the problem can be reduced in any way before continuing:

$$\frac{23}{3} \times \frac{15}{4}$$

Note that one top and one bottom are both divisible by 3. Divide both by 3 before going on:

$$\frac{23}{1} \times \frac{5}{4}$$

Now multiply the top by the top, and the bottom by the bottom, as you always do in multiplying fractions. The result is:

$$\frac{115}{4}$$

Remember the general rule for translating improper fractions into mixed numbers: divide the top by the bottom. The answer is the whole number, and the remainder is the top of the fractional part. This answer translates to $28\frac{3}{4}$.

Cover the answer below with your pad while you exercise the technique on this problem:

$$4\frac{5}{8} \times 3\frac{1}{6}$$

The answer is $14\frac{31}{48}$. You got it by, first, translating $4\frac{5}{8}$ into $\frac{37}{8}$. Then you translated $3\frac{1}{6}$ to $\frac{19}{6}$. This multiplication shows no reduction possibilities, so you multiply top by top and bottom by bottom to get $\frac{703}{48}$. Translate this back to a mixed number by dividing 703 by 48, and produce the final proper answer of $14\frac{31}{48}$.

Dividing by mixed numbers is just the reverse of multiplying. Translate each to an improper fraction, then *invert* the *divider* and multiply as always.

Let's do the last example as a division:

$$4\frac{5}{8} \div 3\frac{1}{6}$$

The two mixed numbers translate to the same improper fractions: $^{37}/_{8}$ and $^{19}/_{6}$. Since this is division, however, we turn the divider upside down and handle it as a multiplication:

$$\frac{37}{8} \times \frac{6}{19}$$

Pause to look for reduction possibilities. The 6 and the 8 are both divisible by 2, so we can simplify the problem a bit to read:

$$\frac{37}{4} \times \frac{3}{19}$$

Multiplying top by top and bottom by bottom, we get the answer $^{111}/_{76}$. This is an improper fraction, but the top is not twice the bottom so we do not divide. We put down a 1 for the whole-number part, and subtract the bottom from the top to find the top of the fractional part. The final answer is $1^{35}/_{76}$.

Most of the fractions and examples in this chapter have been more complex than the ones you normally run into in your number work. This has been entirely on purpose. Learn to handle those in this chapter well, and simpler ones should be easy.

19

SPEED AND EASE IN DECIMALS

FRACTIONS are one way of expressing quantities of less than 1, or more than 1 but not reaching an exact digit. Decimals are another way of doing the same thing.

Of the two, decimals are usually by far the easier and more convenient way to express fractional quantities. If our measuring systems were based on our ten-base counting system (as is the Continental system of meters, grams, litres, and so on) we would perhaps face fractions only a very few times in our lives. But since we have inherited a jumbled group of weights and measures broken down variously into twelfths (feet), sixteenths (pounds), sixtieths (hours), fourths (gallons) and even 5,280th's (miles), we face fractions all the time.

Only in our U. S. money system are we blessed with a commonsense decimal progression. In all our other measurements, we cling to outrageous counting bases.

Even for these, however, decimal fractions are usually accurate enough. They are not as perfect an expression of many quantities as are fractions, which can express any conceivable quantity with exact preciseness, but the difference is so slight as to be meaningless in most cases. In fact, many of the quantities we consider "hard" or "exact" are only approximations to begin with. 5 apples is precisely 5 apples, but

5 inches or 5 pounds is only as (approximately) close to 5 inches or 5 pounds as our measuring equipment can determine at the time.

Actually, these are two completely different types of numbers. One is a precise quantity; the other is a declaration of comparison to an artificial standard such as acres or gallons. Think for a bit about the essential differentness of the two approaches to numbers, for the sake of your number sense.

As far as preciseness of decimals goes, ⅓ is a prime example. There is no decimal equivalent, nor can there ever be. The fraction ⅓ expresses a certain quantity with complete accuracy. The decimal 0.3333333333333333333 approaches ⅓, but it is *not* ⅓. No matter how many 3's you add, you never quite reach ⅓. .33 is accurate to 1 part in 100, however, while .333 is accurate to 1 part in 1,000. For most practical needs, this is more than enough accuracy.

A decimal is a shorthand way of expressing a fraction that has a bottom of 10, 100, 1,000, or some other multiple of 10. We use the decimal point, the little period, to indicate that the digits following it do not express a whole-number quantity, but a fraction whose bottom is a multiple of 10. The number .3 is the same as $^3/_{10}$. 1.3 is the same as $1\,^3/_{10}$. The point tells us when to stop figuring in whole numbers and begin noting the fraction.

The first lesson usually taught in reference to decimals is how to read them properly. Do you read 0.33 as 33 tenths, hundredths, or thousandths? There is a beautifully simple and reliable trick that removes any possible confusion. Merely pretend that the decimal point itself is a 1, followed by as many 0's as there are digits after the point. This imaginary number is the bottom of your fraction.

Thus 0.3 must be $^3/_{10}$, since the "1" (point) is followed by one digit, and 10 is ten. 0.33 must be $^{33}/_{100}$, because there are two digits after the point and two 0's in 100.

See if you can read 0.4567 with this method. The top of the fraction is 4567, of course. The bottom is a 1 followed by four 0's. So the bottom must be 10,000.

If there are any zeros immediately following the point,

count them as digits too in figuring the bottom. 0.03 is $\frac{3}{100}$, not $\frac{3}{10}$, because there are two digits after the point.

A surprising number of people have trouble determining the proper "bottom" to decimal-form fractions, because they never learned this simple trick. For practice in using it, read the following decimals as fractions by saying aloud both the top and bottom of each fraction, just as we might say .3 as $\frac{3}{10}$:

0.67 0.20 0.432 0.5 0.25 0.6478 .004

If you keep firmly in mind the visualization of the point as a 1 followed by as many 0's as there are digits, you should be able to read the above decimals at sight as $\frac{67}{100}$ (hundredths); $\frac{20}{100}$ (this is the same as $\frac{2}{10}$, but it has a slightly different meaning in decimals. That comes later.); $\frac{432}{1,000}$ (thousandths); $\frac{5}{10}$; $\frac{25}{100}$; $\frac{6478}{10,000}$ (ten-thousandths); $\frac{4}{1,000}$.

Mixed Numbers

Expressing mixed numbers (a whole number plus a fraction) is much easier in decimals than it is in fractional form. Whatever part of the number is in front of (to the left of) the point is a whole number. The part to the right of the point expresses the fraction.

So 22.4 is read as "22 and $\frac{4}{10}$ths."

It is often read, too, as "22 point 4." There is nothing wrong with this. It is a short hand way of reading and saying the number, but it does not drive home the actual quantity involved as firmly as does reading the decimal as a full fraction.

1.43 is read as "1 and $\frac{43}{100}$ths."

Read aloud the number 45.67.

If you read it properly as "45 and $\frac{67}{100}$ (hundredths)," go ahead to read the following numbers. Say the full number followed by the fractional part in terms of tenths, hundredths, thousandths, or whatever:

16.4	892.674	3.21	0.45
32.0	100.001	6.08	21.30

If you hesitated over any of these, particularly the one-thousandth or eight-hundredths, it would be a good idea to review the last few pages before going on.

Adding Decimals

There is no trick at all to adding numbers with decimals in them if you keep the basic rule in mind: line up the points. If you were adding 10,342 to 61, you would line up the right-hand ends of these numbers. The point in a number with a decimal fraction is just as clearly and firmly the end of the whole number as is the end when the digits come to a stop there.

Using this rule, set up the following numbers for addition:

65.3
2.13
100.2
.935

Cover the arrangement below with your pad until you have done your part.

Using the points as the ends of the whole numbers, you line up the above numbers for addition like this:

```
 65.3
  2.13
100.2
   .935
```

That is really all there is to it. Elementary, but very important. Once you have lined up the numbers properly, you simply go ahead and add. Ignore the points, except to put a point in your answer in line with the column. Tens carried back across the point as you add behave just as if there were no point there, which is one of the great advantages of using decimals. They enable you to handle the fractional parts right along with your whole numbers, instead of creating them separately.

Having dismissed addition this easily, we can say the same thing about subtraction: keep your points in line, and "borrow" (or slash) across the point as if it were not even there. One example will make this clear:

```
   2 6 8.7 3
 − 7 9.8 6
 ----------
   2̸ 9̸ 9̸.9̸ 7
```

In both addition and subtraction, you can pretend the points are invisible—as long as you line them up, and make sure to put one in your answer in line with the others.

Multiplying Decimals

When you come to multiplying decimals, you do not bother to line up the points because you have another way of placing the point properly in your answer. Refer back to the chapter on no-carry multiplication, if need be, to refresh your understanding of the following rule:

> Add the digits in the two numbers multiplied. Starting with the very left top digit (including the 0 if it is a 0), count this many digits for the answer.

In multiplying decimals, add only the following special qualifier to the general rule:

—to the left of the point.

That "to the left of the point" applies both to the numbers multiplied, and to the answer as well.

It works like this:

```
   3 4.5
 × 2 1.6
 -------
   0 6 9 0
     0 3 4 5
       1̲ 0 7 0
 -------------
     6̲ 4 4̲.2 0
```

How did we place that point in the answer? Each of the two numbers multiplied has two digits to the left of the point. So our answer should have four digits before the point. We start at the very left of the top line with the 0 that does not show up in the final answer but that does have to be counted.

In every other aspect of the problem, we simply ignore the points altogether. You can prove it out by nines-remainders or elevens-remainders, ignoring the points for this purpose too *except* that you start at the point in figuring odd and even digits for an eleven-remainder. If you use continuous subtraction, just keep right on subtracting as you go past the point.

In the example above, you could also have used the classic "point off as many places from the right as there are places to the right of the point in the two numbers multiplied." To rely on this method, however, would rob you of the rapid-estimating nature of no-carry multiplying. Work from left to right instead of right to left, and you can do just as much of any problem as you need to in order to get the accuracy required in that particular situation.

Dividing Decimals

For dividing decimals, we cannot improve on the usual rule: move the point in the divider (if any) all the way to the right. Put a point in your answer as many places to the right of the point in the number divided (if any) as you moved the point in the divider.

If this means adding 0's to the number divided in order to move your point far enough, go ahead and add them.

Here are two examples:

$62.3\,/\overline{48.68}$ becomes $623\,/\overset{\qquad\;\cdot}{\overline{486.8}}$

$4.962\,/\overline{85}$ becomes $4962\,/\overset{\qquad\;\;\cdot}{\overline{85000}}$

Try it yourself. Where will the point in the answer appear for each of the following problems?

$0.6\,/\overline{345.78}$ $\quad$ $22\,/\overline{5.69}$ $\quad$ $4.63\,/\overline{7984.36}$ $\quad$ $78.4\,/\overline{.398}$

Each of these is a little different, but all operate on exactly the same system.

1. The point in the answer will be between the 7 and 8 of the number divided, because we move the point one place to the right.
2. The point in the answer will be directly above the point in the number divided, since there is no point in the divider.
3. The point in the answer will be after the final 6 in the number divided. Moved two places.
4. The point in the answer will be between the 3 and 9 in the number divided. Moved one place.

Other than placing your decimal point properly in the answer, there is no more to dividing with decimal numbers than there is to any division. Once you have determined the right place for the point, simply ignore all the points in the original problem. Your answer will be correct.

Only one other aspect needs special mention. We demonstrated it before, but it should be spelled out too. If you have to move the point in your answer way beyond the end of the number divided, simply do it. Fill in with 0's as needed. For instance:

$$.398\,/\overline{658} \text{ becomes } 398\,/\overline{658000}^{\,\cdot}$$

Decimal Remainders

Depending on the particular problem and the particular field in which your answer will be used, you may work out a division problem that has a remainder in either fractional or decimal form.

The making of a decimal remainder is very simple. It makes no difference how many 0's you add after the last digit to the right of the point, any more than it makes any difference how many 0's you add to the left of a whole number.

00045.2 is the same as 45.2000

There is one special meaning to 0's following the last digit to the right of a decimal point, however, and you should be aware of it. By common agreement, the 0 you place to the right means that the number is accurate to this point.

The number 4.6 might be a rounded-off number anywhere from 4.56 to 4.64. But the number 4.60 means that any rounding off was done beyond the 0. The convention in mathematics goes further, incidentally, and often places a plus or minus sign at the end of a number that has been rounded off, to indicate that it is not a precise quantity.

To make a decimal remainder, then, you simply keep mentally bringing down 0's as long as you have to in order to get an exact answer or the accuracy you need. With your mastery of shorthand division, you do not even have to note the 0's in the number divided; just bring down imaginary ones:

$$\begin{array}{r} 2.5 \\ 4 \; / \overline{\;10} \\ 2 \end{array}$$

If the final division here had not come out even, you would keep bringing down imaginary 0's until you had no remainder, or had as complete an answer as you needed. If you divide 3 into 10, you will never get a complete answer. But at some point you will have as complete an answer as you need.

Converting from Fractions

A fraction, as we have said, is only a special way of writing a division problem. It expresses a specific quantity, but one that (except by decimals) we have no other way of showing with the numbers available than as a division of two known numbers. $\frac{3}{8}$ is the same as $3 \div 8$ or $8 \; / \overline{3}$. The fraction has a different purpose from the division, however; it says, in effect, "this is a quantity," rather than "here is a problem," because for many purposes $\frac{3}{8}$ is more convenient than other expressions of that quantity.

Often, however, you want to convert a fraction to a decimal form. The method is simplicity itself. Simply carry out the implied division, and use a decimal remainder.

To convert $\frac{3}{8}$ to a decimal, for instance, you do this:

```
       .3 7 5
  8 / 3
        6 4
```

The decimal equivalent of $\frac{3}{8}$ is 0.375. In this case, it is an exact equivalent, and it should sound familiar: 375 is one of the basic aliquots.

Now you convert $\frac{6}{7}$ to a decimal. Get out your pad and cover the answer. Express $\frac{6}{7}$ as a decimal accurate to the nearest 10,000th.

Here is how the conversion looks in shorthand division:

```
       .8 5 7 1 4
  7 / 6
        4 5 1 3 2
```

The nearest 10,000th means four places after the point. We worked it out to five places so we could round off, and the last 4 indicates that the rounded-off form is .8571.

Sometimes you find it necessary to convert decimals back to fractions for particular purposes. In some problems, fractions are easier to handle. This, in fact, is part of the basis of the aliquot short cut.

For decimals other than aliquots, the process for converting to a fraction is to write it in fractional form and then see if it can be reduced. The decimal .1 can be written as $\frac{1}{10}$ and .45 can be written as $\frac{45}{100}$.

You reduce this resulting fraction exactly as you reduce any other fraction: divide both top and bottom by any number that will divide both exactly, if there is any. Try reducing the example above, .45.

45 is exactly divisible by 5 or by 9. 100, however, is divisible by 5 but not by 9. Dividing both top and bottom by 5, we reduce $\frac{45}{100}$ to $\frac{9}{20}$. No further reduction is possible.

Convert the following decimals to fractions:

.25 .8125 .625 .96875

The last one, admittedly, is a dilly. But it can be reduced quite substantially. Cover the reductions with your pad until you are satisfied.

Your answers should read $\frac{1}{4}$, $\frac{13}{16}$, $\frac{5}{8}$, and $\frac{31}{32}$.

The next chapter will take up decimals in another and quite special form, percentage. Before going on to that chapter, reflect for a moment or two on the entire decimal method of expressing fractions—and its firm foundation on the point made several times before in this book that each digit decreases in importance by a factor of 10 as it moves each place to the right. This is true right across the decimal point—which is the end of the whole number.

20

HANDLING PERCENTAGES

A PERCENTAGE is merely a two-place decimal without the decimal point shown.

Except that it seems to be the cause of so much general lip-biting, we would dismiss percentages with the above definition. 82% is exactly the same as .82. 6% is no more and no less than .06 (two places, remember). 4½% is .04½, or .045.

A decimal-form fraction with two digits to the right of the point is in hundredths—a "1" followed by as many 0's as there are digits to the right of the point. The term *per cent* comes from the same root as century (a hundred years) and cent (one-hundredth of a dollar): the Latin word for a hundred. Per cent is our contraction of the original *per centum*—per hundred.

So if you say you will pay interest on a loan at the rate of 7% a year, for instance, you are saying that for each 100 parts of the loan you will pay 7 parts a year in interest. If the loan is for $300, you will pay $21 a year; there are 3 100's, and you will pay 7 for each of them. You get precisely the same result if you multiply 300 by .07.

Since we often handle percentages in different ways, let us explore some of the basic relationships and processes involved.

Finding a Percentage of a Number

Finding a percentage of a number is what we just did, and it is the simplest of all percentage calculations. Just multiply the number by the decimal equivalent of the percentage, and you have the answer.

Try one yourself: find 36% of 298.

Here, in no-carry multiplication, is the way you work it out. 36% is, by definition, the same as $^{36}/_{100}$, or .36:

$$\begin{array}{r}
2\ 9\ 8 \\
.3\ 6 \\
\hline
0\ 8\ 9\ 4 \\
1\ 7\ 8\ 8 \\
\hline
0\ \underline{9}\ \underline{6}.2\ 8
\end{array}$$

How do we place the decimal? Remember the decimal rule. The answer has the same number of digits (to the left of the point) as do the two numbers multiplied (to the left of the point). 298 has three places, .36 has none, so the answer has three digits to the left of the point including the first digit of the first partial product, even if it is a 0. The answer is 107.28.

Do one more:

$$\begin{array}{r}
1\ 4.5 \\
8\% \\
\hline
\end{array}$$

Cover the solution with your pad until you have finished this to your satisfaction.

8% is the equivalent of .08, and our solution looks like this:

$$\begin{array}{r}
1\ 4.5 \\
.0\ 8 \\
\hline
0\ \underline{0}.1\ 6
\end{array}$$

Note that there seems to be a spare 0 in the answer. This is to aid the placing of the point in the answer, since the multiplier (.08) has in effect *minus one* places before the point. If we include the 0 in .08 in writing our answer, the correct handling of the point is automatic. We place it two spaces to the right because there are two places to the left of the points in the numbers multiplied.

Finding What Per Cent A Number Is

Often you need to find what per cent one number is of another. You might have, for instance, the two numbers 15 and 75, and be required to express one of them as a percentage of the other.

The important thing is to make very sure which number is which. Do you want to know what per cent 15 is of 75, or what per cent 75 is of 15? It makes a big difference.

Recall at this point that a per cent is only a special way of writing a decimal, and that a decimal is a special form of fraction. So in either of the above cases, you are really being asked to show a fraction in percentage form.

If you want to know what per cent 15 is of 75, you need to convert into decimal (and therefore percentage) form the fraction $^{15}/_{75}$. If you are required to state what per cent 75 is of 15, you again must convert into decimal and percentage form the fraction $^{75}/_{15}$.

Another way of keeping your relationships absolutely straight, in case this conversion does not lock itself memorably in your mind, is that one of the numbers always follows the word *of*. You always ask "what per cent is this number *of* that?" The number *following the "of"* is always the *base*—the base of which you are figuring a percentage—and the *base* is always the *bottom* of the fraction.

You know perfectly well how to convert any fraction to decimal form. You divide the top by the bottom. To convert this decimal fraction to a per cent, move the decimal point two places to the right.

What per cent is 15 of 75?

The fraction to which we want a percentage answer is $^{15}/_{75}$. Using the other key, the number following "of" is 75, and the base is the bottom—again, $^{15}/_{75}$. Now convert:

$$\begin{array}{r} .2 \\ 75\,\overline{)\,15} \end{array}$$

Move the point two places to the right, and we have the answer 20%. 15 is 20% of 75.

Turn the relationship around. What per cent is 75 of 15? Here the fraction expressing the relationship is $^{75}/_{15}$. Or, again, the number following "of" is 15 and therefore the base and the bottom. Divide:

$$15\,\overline{)\,75}\quad = 5.$$

In order to convert this in turn to a percentage, move the point two places to the right—adding 0's as necessary. So 75 is 500% of 15.

500% means that for each 100 parts of the other number, you have 500 parts of this one. Wiping out the 100's, you see that 500% is the same as five times as much.

Try one on your own now. Cover the explanation below with your pad and work out both sides of this relationship:

20 is what per cent of 50?

50 is what per cent of 20?

For the first comparison, the number following the "of," and therefore our base, is 50. The fraction is $^{20}/_{50}$. Dividing by the bottom, we get

$$50\,\overline{)\,20}\quad = .4$$

We move the point two places to the right, and find that 20 is 40% of 50.

Reversing the question, we have a base of 20—the number following the "of." The fraction is $^{50}/_{20}$. The division is

$$\begin{array}{r} 2.5 \\ 20\,\overline{)\,50} \\ 1 \end{array}$$

Again we move the point two places to the right. 50 is 250% of 20.

In these examples, we have not bothered to reduce each fraction to its simplest form before dividing because showing the division with the original numbers in the question seems to make the process clearer. In practice, of course, you would consider these numbers 2 and 5 rather than 20 and 50.

Finding An Unknown Base

One of the most baffling operations in percentage seems to be finding an unknown base. If you have a clear grasp of the relationships, however, it becomes quite easy.

An example of this situation might be the question, "90 is 45% of what?"

We know the number that is a percentage of another. We know the percentage. But we do not know the base.

Let us approach the method through logical conversion of the methods we already understand. Once you know why, you are not likely to forget how.

We have three numbers: 90, 45%, and "what." The number (unknown) following "of" is "what," so "what" is the base.

The fraction, therefore, is

$$\frac{90}{\text{what}}$$

We know the answer to the fraction, but we do not know the fraction itself. In order to convert a fraction to a decimal, and therefore a percentage, we divide the top by the bottom. So we will set up the problem, along with the answer we know:

$$\begin{array}{r} .45 \\ \text{what} \,\big)\overline{\,90\quad} \end{array}$$

Now, if someone asked you, without confusing matters by including words such as percentage and decimals, the question, "What divided into 90 gives the answer .45?" you would answer without a second thought, "Divide .45 into 90 and find out."

Divider multiplied by answer must give number divided. Number divided, divided by the answer, must give the divider.

So we simply divide the number we have by the percentage, and we find the base:

```
       2 0 0.
.4 5 / 9 0
```

Note how the decimal point was moved over, following the rule in the chapter on decimals.

90 is 45% of 200.

The reason we developed this method step by step is to emphasize the logical reasoning behind the general rule:

> To find an unknown base, convert the percentage to a decimal and divide it into the known number.

Reinforce this rule at once by trying another example. 68 is 20% of what?

Convert the percentage into a decimal and divide it into the known number:

```
       3 4 0.
.2 0 / 6 8
         8
```

68 is 20% of 340.

Try one by yourself. Cover up the solution with your pad. 87 is 30% of what?

To find the unknown base, convert the percentage into a decimal and divide it into the known number:

```
          2 9 0.
.3 0 0 / 8 7
         2
```

87 is 30% of 290.

Percentage of Change

Business arithmetic often involves a percentage of change or difference. Rather than asking what per cent 18 is of 360, the business world is more apt to ask, "How much more is 500 than 475?" or, "How much less is 390 than 415?"

Suppose that sales in territory #8 were $350,000 last

year, and are $375,000 this year. What is the percentage of increase?

The first step is to find the raw amount of the difference in plain numbers. It is $25,000, found by subtracting the total last year from the total this year.

Now our problem is, "$25,000 is what per cent of $350,000?"

This is familiar. You did similar problems a few pages ago. The dollar signs and 0's do not change the principle. In fact, you can simplify matters by dropping both the dollar signs and the *same* number of 0's: 25 is what per cent of 350?

Remember your base, the number following "of." The fraction is

$$\frac{25}{350}$$

Work out the division to convert this fraction to decimal form in shorthand division:

```
            .0 7 1
350 / 2 5
        1 5
          2 5
```

The answer is not precise, but we can round it off to 7%. Territory # 8 is 7% ahead of last year.

The general rule, then, is this:

Find the difference, and divide it by the base.

Sometimes the base is the smaller of the two numbers; sometimes it is the larger. After all, sales in territory #8 might have gone down this year. Then the base would be the larger of the two figures.

Do this one on your pad:

Sales last year $320,000
Sales this year $307,200

What is the percentage of decrease?

When we find a percentage of decrease, our base is the

larger number. The difference in sales, by subtraction, is $12,800. Dividing by the base—dropping thousands and dollar signs—we have.

$$320 \overline{) 12.8}^{\;.04}$$

This territory is, unhappily, 4% behind last year in sales.

Note especially that sometimes you figure the percentage of difference on the smaller of two numbers, and sometimes on the larger. The difference, as a percentage, will be larger when based on the smaller number—and smaller when based on the larger number.

The saving grace, perhaps, is that an increase in sales from $100,000 to $150,000 will show up as a 50% increase, while a decline from $150,000 to $100,000 is only 33%!

Now that we have covered decimals and percentage, we are equipped to cover the more common business expressions such as discount and interest and some of the other yardsticks most frequently used in the commercial world.

21
BUSINESS ARITHMETIC

THIS chapter will cover once over lightly the more common business expression involving arithmetic.

The first of these is discount, or mark-down. Retail stores figure the discount they get from the manufacturer or wholesaler with the retail price as the base (book stores, hardware stores, most specialized stores) or—just the opposite in some fields— with the net, discounted price as the base (department stores, chain stores, etc.). When the net price is the base, the store figures mark on or mark up, rather than mark down.

The difference becomes clear in a concrete example.

Mark-down

Suppose a lawn mower retailing for $150 comes to the store with a 30% discount. What is the net price to the store?

The base here is $150. Change the percentage to a decimal and simply multiply. The discount in dollars is .30 times $150, or $45. The net price is $150 minus $45, or $105.

Short cut: The quickest way to figure a net price is not to work out the discount in dollars and then subtract, but to mentally convert the discount into its complement (of 100) and multiply the retail price directly by this. If the retailer gets a 30% discount. then he naturally pays 70% of the retail price.

.70 × $150 gives $105 in one operation, without subtracting.

Try one yourself. A typewriter with a list price of $85 carries a 15% discount to the store. What does the retailer pay for it?

The standard way of doing this is to take .15 of $85, or $12.75, and deduct this from $85 to get a net price of $72.25. The short way is to note that the dealer, in getting a 15% discount, pays 85% of the retail price. So we multiply .85 × $85 and, again, get $72.25 in one operation.

Mark-up

The opposite expression used in many fields is to begin with the net price (the discounted price to the dealer) and arrive at a desired selling price by deciding how much mark-up is required.

A store might have a desired 20% mark-up, for instance. If it buys baby carriages at $30 each net, how much should it sell them for?

Mark-up is figured with the net price as the base, rather than the retail price, so 20% of $30 is $6.00. Adding the cost and the mark-up, the store will sell its baby carriages for $36.

Once again, this can be done without adding, in one operation, by considering that adding 20% to the net price is the same as multiplying the net price by 120%. In this case the short cut is not so effective, however, since you add in the process of multiplying anyway.

Work out a proper selling price for an article that costs $47 and should deliver a 40% mark-up to the store.

For this calculation 40% becomes .4, and .4 × $47 is $18.80. Adding $18.80 to $47, we find a desired retail price of $65.80.

Compound Discounts

Frequently discounts from the retail price are quoted in compound or chain fashion. Toy jobbers (local wholesalers who stock toys and resell them to stores) often buy at discounts such as 50% plus 10%, often called "50 and 10."

This discount is by no means as simple as it looks. It is *not* the sum of 50 and 10; that is, it is not equal to a 60% discount. This is because the second discount is figured on the net price after the first discount, not on the full retail price.

This becomes clear if we start with a $100 item. The 50% discount gives us a first net price of $50. The 10% discount is now applied to the $50, not to the $100, and amounts to $5. This leaves a net-net price of $45. If we had totaled the discounts, we should have figured a net-net price of $40.

The very general 2% cash discount operates in the same way. In order to get their money quickly, most manufacturers allow an extra 2% off the *net* amount of the bill if it is paid by the 10th of the following month.

If our $100 item came to a jobber on such terms, he could (by prompt payment) deduct 2% of the net price. This is 2% of $45, not of $100, so it amounts to 90¢ rather than $2.00. The 2% is important over the total picture (2% can be the profit-margin in some types of business) even if it does not seem spectacular on this $100 item.

So the net result of buying a $100-at-retail toy at a discount of 50% plus 10% plus 2% is that you pay $44.10.

It saves time, in a business in which such discounts prevail, to work out equivalents for the most usual combinations. We have just noted that a discount of 50% plus 10% plus 2% is in effect 55.9% off the retail price.

Turn to your pad and, using 100 as a convenient starting point, work out equivalent one-step discounts for the following compound discounts:

30% plus 5%
40% plus 10%
20% plus 10% plus 5%

The equivalent discounts for these three compound or chain discounts are 33½%, 46%, and 31.6%. Not nearly as generous as they look—which is the reason for quoting them in compound form. They appear to be better than they really are.

Figuring Discounts

A chair retailing for $26 costs the store $18.20. What is the discount percentage?

This is the familiar problem we covered in the chapter on percentage—the process of finding the percentage of difference. The difference here (subtract net from retail) is $7.80. $7.80 is what per cent of $26?

Remember to divide by the base, the number following "of." Our fraction is 7.80/26:

```
          .3
  2 6 / 7.8 0
```

Move the decimal point two places to the right to convert the decimal to a percentage: 30%.

Suppose we want to know the percentage of mark-up in this same case? The net price is now our base, so the fraction is 7.80/18.20:

```
                   .4 0 6
  1 8.2 0 / 7.8 0
              2 2
                2 1 8
```

We see at once that the next digit of the answer will be 5 or more (2 into 10), so we can move the point over to convert to a percentage and round off to 40.7%.

Note how much more the mark-up is as a percentage than is the discount. This is always true, because the base (the net price) is smaller than the retail price.

Break-even

A common expression in many business endeavors is the phrase "break-even point." There are many special applications, but in general the phrase describes the minimum quantity (or volume) required before a product or operation can break even and begin to make a profit.

In tooling up for a new plastic toy, for instance, a manufacturer may have to spend $20,000 in research and die-making costs. If the toy sells for $1.00 retail and he gives the normal 50% plus 10% discount, then he receives 45¢ for each toy. His selling overhead may be 10%, his raw cost of plastic, manufacture, packing and shipping another 10%, and his general company overhead 20%, or a total of 40% of that 45¢ (since the manufacturer figures his volume on *his* sales volume, not the retail price). This leaves 60% of that 45¢ to pay back the cost of getting ready to produce the toy, or 27¢ each. How many toys does he have to sell before he begins to make a profit?

The answer is found by dividing the "contribution" of each sale (27¢) into the "plant account," as it is often called:

```
               6 3 .. .. .. .
.2 7 / 2 0,0 0 0
       1 4 8
         1 1
       1 3 9
         1 2
```

He will have to sell roughly 74,000 of this toy before he recovers his initial investment. Once that investment has been recovered, however, he stands to make 27¢ profit for each toy sold.

A very similar type of calculation is used to determine the break-even point of volume for, say, a grocery store. In any break-even problem, certain assumptions are made about "fixed" costs, such as the plant account for the toy above, or the running expenses of a store, and "variable" costs, or costs that are incurred only when each sale is made.

If all the fixed costs for a certain store were $1,000 a month—including rent, salaries, insurance, etc.—and the average net profit before overhead was 12%, then it is not difficult to calculate how much volume this store must do in order to break even. 12% is the contribution of sales to fixed overhead, or 12¢ on the dollar, so we divide the fixed cost again by the contribution:

```
           8 3 3 3.
.1 2 / 1 0 0 0
       1 1 4
           1 4
```

This store must do over $8,300 a month in sales volume before it can meet its fixed costs. For every dollar above that it does each month, it returns 12¢ profit.

Commission

Salesmen, stockbrokerage houses, insurance agents, and many other companies and people are paid in commission rather than by salary.

Commission is a simple percentage of the gross, or retail price (or net price, depending on the agreement). If a real-estate broker arranges the sale of a house for $20,000 and earns a 5% commission, he gets $1,000.

Commissions vary widely. Salesmen, depending on the field of business, may earn from 1% or 2% to 15% or even more. Stockbrokers work on a sliding scale that goes down as the volume goes up, on the theory that there is about as much paper work in buying or selling $50 worth of stock as there is in buying or selling $100,000 worth. Advertising agencies traditionally get a 15% discount (commission) on the space they buy from magazines or newspapers and the time they buy on radio or television.

As in any percentage situation, you can start with any two known factors and calculate the third, unknown one.

These three cases represent each possible type of unknown. See if you can answer each of them:

A salesman is on 6% commission. He makes a $480 sale. How much commission does he earn by this sale?

Another salesman, on 8% commission, earned $64 one afternoon. How much business did he write in order to get the commission of $64?

A third salesman, on orders totaling $1,300, earned $91 in commissions. What is his commission rate?

Cover the answers with your pad as you work these out.

The first salesman merely has to multiply $480 by .06. He earns $28.80.

The second salesman has to find the base. $64 is 8% of what? As you remember from the chapter on percentage, he determines the unknown base by dividing $64 by .08:

$$\begin{array}{r} 8\ 0\ 0. \\ .0\ 8\ /\overline{6\ 4\ \quad} \end{array}$$

The third salesman also has to divide, but he divides his commission by the base in order to make sure of his rate. The answer is 7%.

Interest

Most of us deal with interest in our personal lives, whether or not we deal very much with it in business. We buy homes almost invariably with a mortgage carrying interest charges. Often we buy automobiles, major appliances, furniture on "time payments" that include interest, whether or not the interest is called that. Sometimes it is called "carrying charges." A bank loan or finance company loan always carries interest charges.

Compound interest is an intriguing subcategory that has little actual utility for most of us. It merely means that the interest is continually added to the principal on which interest is paid, so there is eventually a snowballing effect that can become quite dramatic after a century or two. Except for large interest rates and long periods of time, however, there is little difference in the results.

The interest you receive on your savings account, or the interest you pay on most mortgages, is a "real" interest, figured periodically on the amount of money the bank owes you or you owe the person holding the mortgage.

If you owe $16,000 on a 6% mortgage, the proper charge for one month for the use of this money is $\frac{1}{12}$ of .06, or $\frac{1}{2}$ of 1% (.005), which works out to $80. We use $\frac{1}{12}$ of the interest rate for one month because interest is (unless otherwise stated) figured by the year.

All fair and square. With your mastery of percentages, you should have no trouble with any problem in this area.

But interest, in today's world, has become quite a different thing for most of us. A lender may "prove" to you in black and white that he is charging you 8% interest, yet really can be quite legally gouging you to the extent of 16% or even more. This is so important to almost everybody who borrows money any time in his life that it is worth a page of special explanation.

Hidden Interest

Let us show how the most honest, time-honored, and respectable type of loan from the most inexpensive possible place works: a new-car loan from a bank.

Banks are by far the most reliable and safe places with which to do this kind of business. But when you take out a new-car loan and they say you will pay 6% interest, the reality is that you will pay more than 12%. At a finance company, this could easily go over 24% in real interest charges.

This is why:

When you borrow money and agree to pay interest for the use of it, you properly pay interest on the money while you have it. This is the way mortgages and savings accounts work. Each month (or quarter), the interest on the balance owed is figured and you pay it.

But it does not work this way with consumer loans.

Suppose you go to a bank to borrow, say, about $1,100 to help buy a new car. Your credit standing is good, so the bank says, "Fine." They will charge you only 6% interest, deducted in advance. This means that you sign a note for $1,200, payable in 12 monthly installments. From this $1,200 they now deduct 6% interest for the year it will take you to pay back the loan. 6% of $1,200 is $72, so they give you a check for $1,128 and you buy your car.

The real interest on this loan is more than twice the 6% quoted. Why? For two reasons, First, you *never got* the $1,200 on which you pay the 6%. You got only $1,128. Second, you do not *have* the money for a year at all. You start paying it back the very next month—but the money you pay back the next month has had interest charged for a full year.

Here is how the interest would be charged if you were paying a real 6% on the amount you owe, making payments every month. We chose a $1,200 note to make the figuring easy, since you pay back $100 a month for a year.

MONTH OF LOAN	AMOUNT YOU OWE	PROPER INTEREST FOR THE MONTH AT 6% A YEAR
1	$1128	$ 5.64
2	1028	5.14
3	928	4.64
4	828	4.14
5	728	3.64
6	628	3.14
7	528	2.64
8	428	2.14
9	328	1.64
10	228	1.14
11	128	.64
12	28	.14
		$34.68

The real 6% interest on such a loan totals $34.68. The discounted-in-advance-for-the-full-term arrangement has you pay $72—over *twice* as much.

This illustration is not designed to malign banks. They are the most trustworthy of all such institutions. But if a quoted interest rate at a bank can be so deceptive, imagine what your real charges can become at finance companies when they talk about "only" 8% or 12% a year—discounted in advance.

BIBLIOGRAPHY

AMUSEMENTS IN MATHEMATICS
H. E. Dudeney; Dover Publications, N.Y.

ARITHMETICAL EXCURSIONS
Henry Bowers & Joan E. Bowers; Dover Publications, N.Y.

HIGH-SPEED MATH
Lester Meyers; D. Van Nostrand Company, Inc., N.Y.

HOW TO CALCULATE QUICKLY
Henry Sticker; Dover Publications, N.Y.

THE JAPANESE ABACUS: ITS USE AND THEORY
Takashi Kojima; Charles E. Tuttle Company, Rutland, Vt., and Tokyo

MAGIC WITH FIGURES
A. Frederick Collins; Surrey House, N.Y.

MAGIC HOUSE OF NUMBERS
Irving Adler; The John Day Company, Inc., N.Y.

MATHEMATICS MADE SIMPLE
Abraham Sperling and Monroe Stuart; Doubleday & Co., Inc., Garden City

MATHEMATICIAN'S DELIGHT
W. W. Sawyer; Penguin Books, Ltd., Harmondsworth, England

101 PUZZLES IN THOUGHT AND LOGIC
C. R. Wylie Jr.; Dover Publications, N.Y.

RAPID CALCULATIONS
A. H. Russell; Emerson Books, Inc., N.Y.

SHORT-CUT MATHEMATICS
B. A. Slade, Editor; Nelson-Hall Company, Chicago

THE TRACHTENBERG SPEED SYSTEM OF BASIC MATHEMATICS
Ann Cutler and Rudolph McShane; Doubleday & Co., Inc., Garden City

A CATALOG OF SELECTED
DOVER BOOKS
IN ALL FIELDS OF INTEREST

A CATALOG OF SELECTED DOVER BOOKS IN ALL FIELDS OF INTEREST

CONCERNING THE SPIRITUAL IN ART, Wassily Kandinsky. Pioneering work by father of abstract art. Thoughts on color theory, nature of art. Analysis of earlier masters. 12 illustrations. 80pp. of text. 5⅜ x 8½. 23411-8 Pa. $4.95

ANIMALS: 1,419 Copyright-Free Illustrations of Mammals, Birds, Fish, Insects, etc., Jim Harter (ed.). Clear wood engravings present, in extremely lifelike poses, over 1,000 species of animals. One of the most extensive pictorial sourcebooks of its kind. Captions. Index. 284pp. 9 x 12. 23766-4 Pa. $14.95

CELTIC ART: The Methods of Construction, George Bain. Simple geometric techniques for making Celtic interlacements, spirals, Kells-type initials, animals, humans, etc. Over 500 illustrations. 160pp. 9 x 12. (Available in U.S. only.) 22923-8 Pa. $9.95

AN ATLAS OF ANATOMY FOR ARTISTS, Fritz Schider. Most thorough reference work on art anatomy in the world. Hundreds of illustrations, including selections from works by Vesalius, Leonardo, Goya, Ingres, Michelangelo, others. 593 illustrations. 192pp. 7⅛ x 10¼. 20241-0 Pa. $9.95

CELTIC HAND STROKE-BY-STROKE (Irish Half-Uncial from "The Book of Kells"): An Arthur Baker Calligraphy Manual, Arthur Baker. Complete guide to creating each letter of the alphabet in distinctive Celtic manner. Covers hand position, strokes, pens, inks, paper, more. Illustrated. 48pp. 8¼ x 11. 24336-2 Pa. $3.95

EASY ORIGAMI, John Montroll. Charming collection of 32 projects (hat, cup, pelican, piano, swan, many more) specially designed for the novice origami hobbyist. Clearly illustrated easy-to-follow instructions insure that even beginning papercrafters will achieve successful results. 48pp. 8¼ x 11. 27298-2 Pa. $3.50

THE COMPLETE BOOK OF BIRDHOUSE CONSTRUCTION FOR WOODWORKERS, Scott D. Campbell. Detailed instructions, illustrations, tables. Also data on bird habitat and instinct patterns. Bibliography. 3 tables. 63 illustrations in 15 figures. 48pp. 5¼ x 8½. 24407-5 Pa. $2.50

BLOOMINGDALE'S ILLUSTRATED 1886 CATALOG: Fashions, Dry Goods and Housewares, Bloomingdale Brothers. Famed merchants' extremely rare catalog depicting about 1,700 products: clothing, housewares, firearms, dry goods, jewelry, more. Invaluable for dating, identifying vintage items. Also, copyright-free graphics for artists, designers. Co-published with Henry Ford Museum & Greenfield Village. 160pp. 8¼ x 11. 25780-0 Pa. $12.95

HISTORIC COSTUME IN PICTURES, Braun & Schneider. Over 1,450 costumed figures in clearly detailed engravings–from dawn of civilization to end of 19th century. Captions. Many folk costumes. 256pp. 8⅜ x 11¾. 23150-X Pa. $12.95

STICKLEY CRAFTSMAN FURNITURE CATALOGS, Gustav Stickley and L. & J. G. Stickley. Beautiful, functional furniture in two authentic catalogs from 1910. 594 illustrations, including 277 photos, show settles, rockers, armchairs, reclining chairs, bookcases, desks, tables. 183pp. 6½ x 9¼. 23838-5 Pa. $11.95

AMERICAN LOCOMOTIVES IN HISTORIC PHOTOGRAPHS: 1858 to 1949, Ron Ziel (ed.). A rare collection of 126 meticulously detailed official photographs, called "builder portraits," of American locomotives that majestically chronicle the rise of steam locomotive power in America. Introduction. Detailed captions. xi+ 129pp. 9 x 12. 27393-8 Pa. $13.95

AMERICA'S LIGHTHOUSES: An Illustrated History, Francis Ross Holland, Jr. Delightfully written, profusely illustrated fact-filled survey of over 200 American lighthouses since 1716. History, anecdotes, technological advances, more. 240pp. 8 x 10¾. 25576-X Pa. $12.95

TOWARDS A NEW ARCHITECTURE, Le Corbusier. Pioneering manifesto by founder of "International School." Technical and aesthetic theories, views of industry, economics, relation of form to function, "mass-production split" and much more. Profusely illustrated. 320pp. 6⅛ x 9¼. (Available in U.S. only.) 25023-7 Pa. $10.95

HOW THE OTHER HALF LIVES, Jacob Riis. Famous journalistic record, exposing poverty and degradation of New York slums around 1900, by major social reformer. 100 striking and influential photographs. 233pp. 10 x 7⅞. 22012-5 Pa. $11.95

FRUIT KEY AND TWIG KEY TO TREES AND SHRUBS, William M. Harlow. One of the handiest and most widely used identification aids. Fruit key covers 120 deciduous and evergreen species; twig key 160 deciduous species. Easily used. Over 300 photographs. 126pp. 5⅜ x 8½. 20511-8 Pa. $3.95

COMMON BIRD SONGS, Dr. Donald J. Borror. Songs of 60 most common U.S. birds: robins, sparrows, cardinals, bluejays, finches, more–arranged in order of increasing complexity. Up to 9 variations of songs of each species. Cassette and manual 99911-4 $8.95

ORCHIDS AS HOUSE PLANTS, Rebecca Tyson Northen. Grow cattleyas and many other kinds of orchids–in a window, in a case, or under artificial light. 63 illustrations. 148pp. 5⅜ x 8½. 23261-1 Pa. $7.95

MONSTER MAZES, Dave Phillips. Masterful mazes at four levels of difficulty. Avoid deadly perils and evil creatures to find magical treasures. Solutions for all 32 exciting illustrated puzzles. 48pp. 8¼ x 11. 26005-4 Pa. $2.95

MOZART'S DON GIOVANNI (DOVER OPERA LIBRETTO SERIES), Wolfgang Amadeus Mozart. Introduced and translated by Ellen H. Bleiler. Standard Italian libretto, with complete English translation. Convenient and thoroughly portable–an ideal companion for reading along with a recording or the performance itself. Introduction. List of characters. Plot summary. 121pp. 5¼ x 8½. 24944-1 Pa. $3.95

TECHNICAL MANUAL AND DICTIONARY OF CLASSICAL BALLET, Gail Grant. Defines, explains, comments on steps, movements, poses and concepts. 15-page pictorial section. Basic book for student, viewer. 127pp. 5⅜ x 8½. 21843-0 Pa. $4.95

MY BONDAGE AND MY FREEDOM, Frederick Douglass. Born a slave, Douglass became outspoken force in antislavery movement. The best of Douglass' autobiographies. Graphic description of slave life. 464pp. 5⅜ x 8½. 22457-0 Pa. $8.95

FOLLOWING THE EQUATOR: A Journey Around the World, Mark Twain. Fascinating humorous account of 1897 voyage to Hawaii, Australia, India, New Zealand, etc. Ironic, bemused reports on peoples, customs, climate, flora and fauna, politics, much more. 197 illustrations. 720pp. 5⅜ x 8½. 26113-1 Pa. $15.95

THE PEOPLE CALLED SHAKERS, Edward D. Andrews. Definitive study of Shakers: origins, beliefs, practices, dances, social organization, furniture and crafts, etc. 33 illustrations. 351pp. 5⅜ x 8½. 21081-2 Pa. $12.95

THE MYTHS OF GREECE AND ROME, H. A. Guerber. A classic of mythology, generously illustrated, long prized for its simple, graphic, accurate retelling of the principal myths of Greece and Rome, and for its commentary on their origins and significance. With 64 illustrations by Michelangelo, Raphael, Titian, Rubens, Canova, Bernini and others. 480pp. 5⅜ x 8½. 27584-1 Pa. $10.95

PSYCHOLOGY OF MUSIC, Carl E. Seashore. Classic work discusses music as a medium from psychological viewpoint. Clear treatment of physical acoustics, auditory apparatus, sound perception, development of musical skills, nature of musical feeling, host of other topics. 88 figures. 408pp. 5⅜ x 8½. 21851-1 Pa. $11.95

THE PHILOSOPHY OF HISTORY, Georg W. Hegel. Great classic of Western thought develops concept that history is not chance but rational process, the evolution of freedom. 457pp. 5⅜ x 8½. 20112-0 Pa. $9.95

THE BOOK OF TEA, Kakuzo Okakura. Minor classic of the Orient: entertaining, charming explanation, interpretation of traditional Japanese culture in terms of tea ceremony. 94pp. 5⅜ x 8½. 20070-1 Pa. $3.95

LIFE IN ANCIENT EGYPT, Adolf Erman. Fullest, most thorough, detailed older account with much not in more recent books, domestic life, religion, magic, medicine, commerce, much more. Many illustrations reproduce tomb paintings, carvings, hieroglyphs, etc. 597pp. 5⅜ x 8½. 22632-8 Pa. $12.95

SUNDIALS, Their Theory and Construction, Albert Waugh. Far and away the best, most thorough coverage of ideas, mathematics concerned, types, construction, adjusting anywhere. Simple, nontechnical treatment allows even children to build several of these dials. Over 100 illustrations. 230pp. 5⅜ x 8½. 22947-5 Pa. $8.95

THEORETICAL HYDRODYNAMICS, L. M. Milne-Thomson. Classic exposition of the mathematical theory of fluid motion, applicable to both hydrodynamics and aerodynamics. Over 600 exercises. 768pp. 6⅛ x 9¼. 68970-0 Pa. $20.95

SONGS OF EXPERIENCE: Facsimile Reproduction with 26 Plates in Full Color, William Blake. 26 full-color plates from a rare 1826 edition. Includes "The Tyger," "London," "Holy Thursday," and other poems. Printed text of poems. 48pp. 5¼ x 7. 24636-1 Pa. $4.95

OLD-TIME VIGNETTES IN FULL COLOR, Carol Belanger Grafton (ed.). Over 390 charming, often sentimental illustrations, selected from archives of Victorian graphics–pretty women posing, children playing, food, flowers, kittens and puppies, smiling cherubs, birds and butterflies, much more. All copyright-free. 48pp. 9¼ x 12¼. 27269-9 Pa. $9.95

LITTLE BOOK OF EARLY AMERICAN CRAFTS AND TRADES, Peter Stockham (ed.). 1807 children's book explains crafts and trades: baker, hatter, cooper, potter, and many others. 23 copperplate illustrations. 140pp. 4⅝ x 6.
23336-7 Pa. $4.95

VICTORIAN FASHIONS AND COSTUMES FROM HARPER'S BAZAR, 1867–1898, Stella Blum (ed.). Day costumes, evening wear, sports clothes, shoes, hats, other accessories in over 1,000 detailed engravings. 320pp. 9⅜ x 12¼.
22990-4 Pa. $16.95

GUSTAV STICKLEY, THE CRAFTSMAN, Mary Ann Smith. Superb study surveys broad scope of Stickley's achievement, especially in architecture. Design philosophy, rise and fall of the Craftsman empire, descriptions and floor plans for many Craftsman houses, more. 86 black-and-white halftones. 31 line illustrations. Introduction 208pp. 6½ x 9¼. 27210-9 Pa. $9.95

THE LONG ISLAND RAIL ROAD IN EARLY PHOTOGRAPHS, Ron Ziel. Over 220 rare photos, informative text document origin (1844) and development of rail service on Long Island. Vintage views of early trains, locomotives, stations, passengers, crews, much more. Captions. 8⅞ x 11¾. 26301-0 Pa. $14.95

VOYAGE OF THE LIBERDADE, Joshua Slocum. Great 19th-century mariner's thrilling, first-hand account of the wreck of his ship off South America, the 35-foot boat he built from the wreckage, and its remarkable voyage home. 128pp. 5⅜ x 8½.
40022-0 Pa. $5.95

TEN BOOKS ON ARCHITECTURE, Vitruvius. The most important book ever written on architecture. Early Roman aesthetics, technology, classical orders, site selection, all other aspects. Morgan translation. 331pp. 5⅜ x 8½. 20645-9 Pa. $9.95

THE HUMAN FIGURE IN MOTION, Eadweard Muybridge. More than 4,500 stopped-action photos, in action series, showing undraped men, women, children jumping, lying down, throwing, sitting, wrestling, carrying, etc. 390pp. 7⅞ x 10⅝.
20204-6 Clothbd. $29.95

TREES OF THE EASTERN AND CENTRAL UNITED STATES AND CANADA, William M. Harlow. Best one-volume guide to 140 trees. Full descriptions, woodlore, range, etc. Over 600 illustrations. Handy size. 288pp. 4½ x 6⅜.
20395-6 Pa. $6.95

SONGS OF WESTERN BIRDS, Dr. Donald J. Borror. Complete song and call repertoire of 60 western species, including flycatchers, juncoes, cactus wrens, many more–includes fully illustrated booklet. Cassette and manual 99913-0 $8.95

GROWING AND USING HERBS AND SPICES, Milo Miloradovich. Versatile handbook provides all the information needed for cultivation and use of all the herbs and spices available in North America. 4 illustrations. Index. Glossary. 236pp. 5⅜ x 8½.
25058-X Pa. $7.95

BIG BOOK OF MAZES AND LABYRINTHS, Walter Shepherd. 50 mazes and labyrinths in all–classical, solid, ripple, and more–in one great volume. Perfect inexpensive puzzler for clever youngsters. Full solutions. 112pp. 8¼ x 11.
22951-3 Pa. $5.95

PIANO TUNING, J. Cree Fischer. Clearest, best book for beginner, amateur. Simple repairs, raising dropped notes, tuning by easy method of flattened fifths. No previous skills needed. 4 illustrations. 201pp. 5⅜ x 8½. 23267-0 Pa. $6.95

HINTS TO SINGERS, Lillian Nordica. Selecting the right teacher, developing confidence, overcoming stage fright, and many other important skills receive thoughtful discussion in this indispensible guide, written by a world-famous diva of four decades' experience. 96pp. 5 3/8 x 8 1/2. 40094-8 Pa. $4.95

THE COMPLETE NONSENSE OF EDWARD LEAR, Edward Lear. All nonsense limericks, zany alphabets, Owl and Pussycat, songs, nonsense botany, etc., illustrated by Lear. Total of 320pp. 5⅜ x 8½. (Available in U.S. only.) 20167-8 Pa. $7.95

VICTORIAN PARLOUR POETRY: An Annotated Anthology, Michael R. Turner. 117 gems by Longfellow, Tennyson, Browning, many lesser-known poets. "The Village Blacksmith," "Curfew Must Not Ring Tonight," "Only a Baby Small," dozens more, often difficult to find elsewhere. Index of poets, titles, first lines. xxiii + 325pp. 5⅝ x 8¼. 27044-0 Pa. $12.95

DUBLINERS, James Joyce. Fifteen stories offer vivid, tightly focused observations of the lives of Dublin's poorer classes. At least one, "The Dead," is considered a masterpiece. Reprinted complete and unabridged from standard edition. 160pp. 5 3/16 x 8¼. 26870-5 Pa. $1.50

GREAT WEIRD TALES: 14 Stories by Lovecraft, Blackwood, Machen and Others, S. T. Joshi (ed.). 14 spellbinding tales, including "The Sin Eater," by Fiona McLeod, "The Eye Above the Mantel," by Frank Belknap Long, as well as renowned works by R. H. Barlow, Lord Dunsany, Arthur Machen, W. C. Morrow and eight other masters of the genre. 256pp. 5⅜ x 8½. (Available in U.S. only.) 40436-6 Pa. $8.95

THE BOOK OF THE SACRED MAGIC OF ABRAMELIN THE MAGE, translated by S. MacGregor Mathers. Medieval manuscript of ceremonial magic. Basic document in Aleister Crowley, Golden Dawn groups. 268pp. 5⅜ x 8½. 23211-5 Pa. $9.95

NEW RUSSIAN-ENGLISH AND ENGLISH-RUSSIAN DICTIONARY, M. A. O'Brien. This is a remarkably handy Russian dictionary, containing a surprising amount of information, including over 70,000 entries. 366pp. 4½ x 6⅛. 20208-9 Pa. $10.95

HISTORIC HOMES OF THE AMERICAN PRESIDENTS, Second, Revised Edition, Irvin Haas. A traveler's guide to American Presidential homes, most open to the public, depicting and describing homes occupied by every American President from George Washington to George Bush. With visiting hours, admission charges, travel routes. 175 photographs. Index. 160pp. 8¼ x 11. 26751-2 Pa. $13.95

NEW YORK IN THE FORTIES, Andreas Feininger. 162 brilliant photographs by the well-known photographer, formerly with *Life* magazine. Commuters, shoppers, Times Square at night, much else from city at its peak. Captions by John von Hartz. 181pp. 9¼ x 10¾. 23585-8 Pa. $13.95

INDIAN SIGN LANGUAGE, William Tomkins. Over 525 signs developed by Sioux and other tribes. Written instructions and diagrams. Also 290 pictographs. 111pp. 6⅛ x 9¼. 22029-X Pa. $3.95

THE WIT AND HUMOR OF OSCAR WILDE, Alvin Redman (ed.). More than 1,000 ripostes, paradoxes, wisecracks: Work is the curse of the drinking classes; I can resist everything except temptation; etc. 258pp. 5⅜ x 8½. 20602-5 Pa. $6.95

SHAKESPEARE LEXICON AND QUOTATION DICTIONARY, Alexander Schmidt. Full definitions, locations, shades of meaning in every word in plays and poems. More than 50,000 exact quotations. 1,485pp. 6½ x 9¼. 2-vol. set.
Vol. 1: 22726-X Pa. $17.95
Vol. 2: 22727-8 Pa. $17.95

SELECTED POEMS, Emily Dickinson. Over 100 best-known, best-loved poems by one of America's foremost poets, reprinted from authoritative early editions. No comparable edition at this price. Index of first lines. 64pp. 5 3/16 x 8¼.
26466-1 Pa. $1.00

THE INSIDIOUS DR. FU-MANCHU, Sax Rohmer. The first of the popular mystery series introduces a pair of English detectives to their archnemesis, the diabolical Dr. Fu-Manchu. Flavorful atmosphere, fast-paced action, and colorful characters enliven this classic of the genre. 208pp. 5 3/16 x 8¼. 29898-1 Pa. $2.00

THE MALLEUS MALEFICARUM OF KRAMER AND SPRENGER, translated by Montague Summers. Full text of most important witchhunter's "bible," used by both Catholics and Protestants. 278pp. 6⅝ x 10. 22802-9 Pa. $12.95

SPANISH STORIES/CUENTOS ESPAÑOLES: A Dual-Language Book, Angel Flores (ed.). Unique format offers 13 great stories in Spanish by Cervantes, Borges, others. Faithful English translations on facing pages. 352pp. 5⅜ x 8½.
25399-6 Pa. $9.95

GARDEN CITY, LONG ISLAND, IN EARLY PHOTOGRAPHS, 1869–1919, Mildred H. Smith. Handsome treasury of 118 vintage pictures, accompanied by carefully researched captions, document the Garden City Hotel fire (1899), the Vanderbilt Cup Race (1908), the first airmail flight departing from the Nassau Boulevard Aerodrome (1911), and much more. 96pp. 8 7/8 x 11 3/4. 40669-5 Pa. $12.95

OLD QUEENS, N.Y., IN EARLY PHOTOGRAPHS, Vincent F. Seyfried and William Asadorian. Over 160 rare photographs of Maspeth, Jamaica, Jackson Heights, and other areas. Vintage views of DeWitt Clinton mansion, 1939 World's Fair and more. Captions. 192pp. 8⅞ x 11. 26358-4 Pa. $14.95

CAPTURED BY THE INDIANS: 15 Firsthand Accounts, 1750-1870, Frederick Drimmer. Astounding true historical accounts of grisly torture, bloody conflicts, relentless pursuits, miraculous escapes and more, by people who lived to tell the tale. 384pp. 5⅜ x 8½. 24901-8 Pa. $9.95

THE WORLD'S GREAT SPEECHES (Fourth Enlarged Edition), Lewis Copeland, Lawrence W. Lamm, and Stephen J. McKenna. Nearly 300 speeches provide public speakers with a wealth of updated quotes and inspiration–from Pericles' funeral oration and William Jennings Bryan's "Cross of Gold Speech" to Malcolm X's powerful words on the Black Revolution and Earl of Spenser's tribute to his sister, Diana, Princess of Wales. 944pp. 5⅜ x 8⅜. 40903-1 Pa. $15.95

THE BOOK OF THE SWORD, Sir Richard F. Burton. Great Victorian scholar/adventurer's eloquent, erudite history of the "queen of weapons"–from prehistory to early Roman Empire. Evolution and development of early swords, variations (sabre, broadsword, cutlass, scimitar, etc.), much more. 336pp. 6⅛ x 9¼.
25434-8 Pa. $9.95

AUTOBIOGRAPHY: The Story of My Experiments with Truth, Mohandas K. Gandhi. Boyhood, legal studies, purification, the growth of the Satyagraha (nonviolent protest) movement. Critical, inspiring work of the man responsible for the freedom of India. 480pp. 5⅜ x 8½. (Available in U.S. only.) 24593-4 Pa. $9.95

CELTIC MYTHS AND LEGENDS, T. W. Rolleston. Masterful retelling of Irish and Welsh stories and tales. Cuchulain, King Arthur, Deirdre, the Grail, many more. First paperback edition. 58 full-page illustrations. 512pp. 5⅜ x 8½. 26507-2 Pa. $9.95

THE PRINCIPLES OF PSYCHOLOGY, William James. Famous long course complete, unabridged. Stream of thought, time perception, memory, experimental methods; great work decades ahead of its time. 94 figures. 1,391pp. 5⅜ x 8½. 2-vol. set.
Vol. I: 20381-6 Pa. $14.95
Vol. II: 20382-4 Pa. $16.95

THE WORLD AS WILL AND REPRESENTATION, Arthur Schopenhauer. Definitive English translation of Schopenhauer's life work, correcting more than 1,000 errors, omissions in earlier translations. Translated by E. F. J. Payne. Total of 1,269pp. 5⅜ x 8½. 2-vol. set. Vol. 1: 21761-2 Pa. $12.95
Vol. 2: 21762-0 Pa. $12.95

MAGIC AND MYSTERY IN TIBET, Madame Alexandra David-Neel. Experiences among lamas, magicians, sages, sorcerers, Bonpa wizards. A true psychic discovery. 32 illustrations. 321pp. 5⅜ x 8½. (Available in U.S. only.) 22682-4 Pa. $9.95

THE EGYPTIAN BOOK OF THE DEAD, E. A. Wallis Budge. Complete reproduction of Ani's papyrus, finest ever found. Full hieroglyphic text, interlinear transliteration, word-for-word translation, smooth translation. 533pp. 6½ x 9¼.
21866-X Pa. $12.95

MATHEMATICS FOR THE NONMATHEMATICIAN, Morris Kline. Detailed, college-level treatment of mathematics in cultural and historical context, with numerous exercises. Recommended Reading Lists. Tables. Numerous figures. 641pp. 5⅜ x 8½.
24823-2 Pa. $11.95

PROBABILISTIC METHODS IN THE THEORY OF STRUCTURES, Isaac Elishakoff. Well-written introduction covers the elements of the theory of probability from two or more random variables, the reliability of such multivariable structures, the theory of random function, Monte Carlo methods of treating problems incapable of exact solution, and more. Examples. 502pp. 5⅜ x 8½. 40691-1 Pa. $16.95

THE RIME OF THE ANCIENT MARINER, Gustave Doré, S. T. Coleridge. Doré's finest work; 34 plates capture moods, subtleties of poem. Flawless full-size reproductions printed on facing pages with authoritative text of poem. "Beautiful. Simply beautiful."–*Publisher's Weekly.* 77pp. 9¼ x 12. 22305-1 Pa. $7.95

NORTH AMERICAN INDIAN DESIGNS FOR ARTISTS AND CRAFTSPEOPLE, Eva Wilson. Over 360 authentic copyright-free designs adapted from Navajo blankets, Hopi pottery, Sioux buffalo hides, more. Geometrics, symbolic figures, plant and animal motifs, etc. 128pp. 8⅜ x 11. (Not for sale in the United Kingdom.) 25341-4 Pa. $9.95

SCULPTURE: Principles and Practice, Louis Slobodkin. Step-by-step approach to clay, plaster, metals, stone; classical and modern. 253 drawings, photos. 255pp. 8⅛ x 11.
22960-2 Pa. $11.95